Handbook of
SAS® DATA Step Programming

Handbook of
SAS® DATA Step Programming

Arthur Li

CRC Press
Taylor & Francis Group
Boca Raton London New York

CRC Press is an imprint of the
Taylor & Francis Group, an **informa** business

A CHAPMAN & HALL BOOK

CRC Press
Taylor & Francis Group
6000 Broken Sound Parkway NW, Suite 300
Boca Raton, FL 33487-2742

© 2013 by Taylor & Francis Group, LLC
CRC Press is an imprint of Taylor & Francis Group, an Informa business

No claim to original U.S. Government works

Printed on acid-free paper
Version Date: 20130226

Printed and bound in Great Britain by TJ International, Padstow, Cornwall

International Standard Book Number-13: 978-1-4665-5238-8 (Hardback)

Visit the Taylor & Francis Web site at
http://www.taylorandfrancis.com

and the CRC Press Web site at
http://www.crcpress.com

To Dave

Contents

Preface xiii
The Author xix
Acknowledgments xxi

1. Introduction to SAS® 1
  1.1   SAS Program and Language 1
  1.2   Reading Data into SAS 2
        1.2.1   The SAS Data Set and SAS Library 2
        1.2.2   Reading a SAS Data Set 3
        1.2.3   Reading a Raw Data File with Fixed Fields 6
        1.2.4   Reading Data Entered Directly into the Program 8
  1.3   Creating and Modifying Variables 9
        1.3.1   The Assignment Statement and SAS Expression 9
        1.3.2   Creating Variables Conditionally 12
  1.4   Base SAS Procedures 13
        1.4.1   Common Statements in SAS Procedures: TITLE, BY,
               and WHERE Statements 13
        1.4.2   The CONTENTS Procedure 14
        1.4.3   The SORT Procedure 16
        1.4.4   The PRINT Procedure 16
        1.4.5   The MEANS Procedure 18
        1.4.6   The FREQ Procedure 20
  1.5   Subsetting Data by Selecting Variables 24
        1.5.1   Selecting Variables with the KEEP= Data Set Option
               or KEEP Statement 25
        1.5.2   Selecting Variables with the DROP= Data Set Option
               or DROP Statement 27
        1.5.3   Where to Specify the DROP= and KEEP= Data Set
               Options and DROP/KEEP Statements 27
  1.6   Changing the Appearance of Data 28
        1.6.1   Labeling Variables 29
        1.6.2   Formatting Variable Values Using SAS FORMATS 31
  Exercises 33

2. Creating Variables Conditionally 35
  2.1   The IF-THEN/ELSE Statement 35
        2.1.1   Steps for Creating a Variable 35
        2.1.2   Handling Missing Values When Creating Variables 37
        2.1.3   TRUE and FALSE: Logical Expressions 39

  2.1.4   The LENGTH Attribute 41
  2.1.5   DO Group...43
2.2   Executing One of Several Statements......................................45
  2.2.1   Multiple IF-THEN/ELSE Statements.................................45
  2.2.2   Executing Statements Using the SELECT Group 48
2.3   Modifying the IF-THEN/ELSE Statement with
        the Assignment Statement...52
Exercises...54

3. Understanding How the DATA Step Works 55
3.1   DATA Step Processing Overview 55
  3.1.1   DATA Step Compilation Phase..57
  3.1.2   DATA Step Execution Phase...58
  3.1.3   The Importance of the OUTPUT Statement......................61
  3.1.4   The Difference between Reading a Raw Data Set
          and a SAS Data Set..61
3.2   Retaining the Value of Newly Created Variables...........................62
  3.2.1   The RETAIN Statement...62
  3.2.2   The SUM Statement...64
3.3   Conditional Processing in the DATA Step 66
  3.3.1   The Subsetting IF Statement 66
  3.3.2   Detecting the End of a Data Set by Using
          the END= Option..68
  3.3.3   Restructuring Data Sets from Wide Format to Long
          Format..68
3.4   Debugging Techniques 70
  3.4.1   Using the PUT Statement to Observe the Contents
          of the PDV 70
  3.4.2   Using the DATA Step Debugger 74
Exercises...76

4. BY-Group Processing in the DATA Step 79
4.1   Introduction to BY-Group Processing..79
  4.1.1   The FIRST.VARIABLE and the LAST.VARIABLE..............79
  4.1.2   The Execution Phase of BY-Group Processing 81
4.2   Applications Utilizing BY-Group Processing 85
  4.2.1   Calculating Mean Score within Each BY Group 87
  4.2.2   Creating Data Sets with Duplicate or Non-Duplicate
          Observations...88
  4.2.3   Obtaining the Most Recent Non-Missing Data
          within Each BY Group 89
  4.2.4   Restructuring Data Sets from Long Format to
          Wide Format 91
Exercises...92

5. Writing Loops in the DATA Step 95
5.1 Implicit and Explicit Loops 95
5.1.1 Implicit Loops 95
5.1.2 Explicit Loops 96
5.1.3 Nested Loops 102
5.1.4 Combining Implicit and Explicit Loops 103
5.2 Utilizing Loops to Create Samples 103
5.2.1 Direct-Access Mode 104
5.2.2 Creating a Systematic Sample 105
5.2.3 Creating a Random Sample with Replacement 106
5.2.4 Creating a Random Sample without Replacement 108
5.3 Using Looping to Read a List of External Files 110
5.3.1 Using an Iterative DO Loop to Read an External File 110
5.3.2 Using an Iterative DO-Loop to Read Multiple
External Files 111
Exercises 117

6. Array Processing 121
6.1 Introduction to Array Processing 121
6.1.1 Situations for Utilizing Array Processing 121
6.1.2 Defining and Referencing One-Dimensional Arrays 123
6.1.3 Compilation and Execution Phases for Array
Processing 125
6.2 Functions and Operators Related to Array Processing 126
6.2.1 The DIM, HBOUND, and LBOUND Functions 126
6.2.2 Using the IN and OF Operator with an Array 129
6.3 Some Array Applications 130
6.3.1 Creating a Group of Variables by Using Arrays 130
6.3.2 Calculating Products of Multiple Variables 131
6.3.3 Restructuring Data Sets Using One-Dimensional
Arrays 132
6.4 Applications That Use Multi-Dimensional Arrays 133
6.4.1 Calculating Average SBP for Pre- and Post-Treatment 133
6.4.2 Restructuring Data Sets by Using
a Multi-Dimensional Array 135
Exercises 136

7. Combining Data Sets 139
7.1 Vertically Combining Data Sets 139
7.1.1 Concatenating Data Sets 139
7.1.2 Interleaving Data Sets 142
7.2 Horizontally Combining Data Sets 143
7.2.1 One-to-One Reading 143
7.2.2 One-to-One Merging 146

    7.2.3   Match-Merging 147
    7.2.4   Updating Data Sets 151
  Exercises 152

8. Data Input and Output 155
  8.1   Introduction to Reading and Writing Text Files 155
    8.1.1   Steps for Reading Text Files 155
    8.1.2   Steps for Writing Text Files 157
    8.1.3   Data Informat 157
    8.1.4   Data Format 158
    8.1.5   SAS Date and Time Values 158
  8.2   Reading Text Files 160
    8.2.1   Column Input 161
    8.2.2   Formatted Input 162
    8.2.3   List Input 164
    8.2.4   Modified List Input 168
    8.2.5   Mixed Input 170
    8.2.6   Creating Observations by Using the Line
               Pointer-Controls 171
    8.2.7   Creating Observations by Using Line-Hold Specifiers 172
  8.3   Creating Text Files 175
    8.3.1   Column Output 175
    8.3.2   Formatted Output 176
    8.3.3   List Output 177
  Exercises 177

9. Data Step Functions 181
  9.1   Introduction to Functions and CALL Routines 181
    9.1.1   Functions 181
    9.1.2   CALL Routines 182
    9.1.3   Categories of Functions and CALL Routines 184
  9.2   Date and Time Functions 185
    9.2.1   Creating Date and Time Values 185
    9.2.2   Extracting Components from Date and Time Values 187
    9.2.3   Date and Time Interval Functions 188
  9.3   Character Functions 190
    9.3.1   Functions for Changing Character Cases 190
    9.3.2   Functions for Concatenating Character Strings 191
    9.3.3   Functions for Searching, Exacting, and Replacing
               Character Strings 194
  9.4   Functions for Converting Variable Types 198
    9.4.1   The INPUT Function 198
    9.4.2   The PUT Function 201
  Exercises 203

10. Useful SAS® Procedures..205

   10.1   Using the SORT Procedure to Eliminate Duplicate
           Observations...205
         10.1.1 Eliminating Observations with Duplicate BY Values......205
         10.1.2 Eliminating Duplicate Observations 207
   10.2   Using the COMPARE Procedure to Compare the Contents
           of Two Data Sets...208
         10.2.1 Information Provided from PROC COMPARE 209
         10.2.2 Comparing Observations with Common ID Values.......212
   10.3   Restructuring Data Sets Using the TRANSPOSE Procedure 215
         10.3.1 Transposing an Entire Data Set 216
         10.3.2 Introduction to Transposing BY Groups 219
         10.3.3 Where the ID Statement Does Not Work for
                Transposing BY Groups..220
         10.3.4 Where the ID Statement Is Essential for Transposing
                BY Groups 221
         10.3.5 Handling Duplicated Observations Using
                the LET Option..222
         10.3.6 Situations Requiring Two or More Transpositions.........224
   10.4   Creating the User-Defined Format Using the FORMAT
           Procedure 227
         10.4.1 Creating User-Defined Formats......................................228
         10.4.2 Retrieving User-Defined Formats 233
         10.4.3 Creating Variables by Using User-Defined Formats.......235
   10.5   Using the OPTIONS Procedure to Modify SAS System
           Options 236
   Exercises...239

References 241

Index..243

Preface

A common statistical programmer's task generally begins with reading one or more raw data sets into a statistical software and performing data management and manipulation, such as checking and modifying values of variables to ensure they satisfy the analytic quality, transposing data into a desired shape for a certain type of statistical analysis, merging multiple data sets by common variables, etc. Once the data has been assured of possessing the desired quality, the transformed data is ready for statistical analysis by using procedures that are provided by the statistical software. Data manipulation is an essential step to obtaining a reliable, statistical, analytical result because an analytical result that is based on unreliable data is not trustworthy; this is often referred to as "garbage in, garbage out." Successfully creating reliable data solely depends upon writing an accurate computer program.

In SAS®, data manipulation and management are mostly performed in the DATA steps; conducting statistical analysis and creating reports is carried out by using SAS procedures (PROC steps). A computer program that is used to perform data manipulation and analysis consists of a series of DATA or PROC steps or both and is written within the SAS programming language. The DATA and PROC steps consist of a series of SAS statements that are created by following the SAS language syntax. A PROC step is often easy to write because it is purely syntax driven; therefore, simply knowing how to follow the syntax from the documentation will sometimes be sufficient to help you accomplish the task successfully. On the other hand, in order to write an accomplished program in the DATA step, a programmer must be able to understand programming logic and to know how to implement and even create his or her own programming algorithm.

The focus of this book is not about learning statistical procedures but rather learning how best to manage and manipulate data by using the DATA step. Beginning programmers often tend to focus on learning syntax without focusing on programming logic and algorithms, which often results in common problems when they create a SAS data set. For example, the data set that they created is not what they originally intended to create—that is, there are more or less observations than intended or the value of the newly created variable was not retained correctly. These types of mistakes are most commonly committed because programming novices don't understand fundamental and unique SAS programming concepts, such as understanding the compilation and execution phases of the DATA step, what happens in the program data vector (PDV) during the DATA step execution, etc. This book will provide insight to readers that simply learning syntax will not solve all the problems that they'll encounter; instead, they need to understand SAS processing in order to be successful programmers.

Another common problem novice programmers face is a lack of programming strategies. Therefore, when SAS programming novices encounter a new programming task, often they don't know where to begin and what steps will be involved in solving the problem. Most of the examples in this book begin with discussing the strategies and steps for solving the problems, then providing a solution, and in the end, providing a more detailed explanation for the solution.

An Overview of SAS Software

SAS was originally the acronym for Statistical Analysis System, which is an integrated software system that utilizes fourth-generation programming language to perform tasks like data management, report writing, statistical analysis, data warehousing, and application development. SAS has undergone various upgrades to its software system over the years. This book was written using Version 9.2.

The core component of the SAS system is Base SAS software, which consists of different modules such as DATA steps, SAS Base procedures, SAS macro facility, and Output Delivery System (ODS). Among the modules in Base SAS, this book covers the DATA step and some of the SAS Base procedures that relate to data management.

SAS provides multiple methods for starting and running your SAS program, which depends upon your operating system. The common method for most SAS users is to utilize the SAS Windowing Environment. Alternatively, you can also run your program by using an interactive or noninteractive line mode, as well as batch mode. Please refer to SAS documentation for these alternative methods.

SAS Windowing Environment (illustrated in Figure 0.1) consists of five windows where you can create and edit your SAS program and manage your SAS files, which includes Program Editor, Log, Output, Result, and Explorer windows. To create a SAS program in the SAS Windowing Environment, you can use the Program Editor window to write your SAS code. You can either submit your entire program at once or highlight only part of the program that you wish to run by clicking on the "Submit" button on the tool bar or by selecting "Run ◊ Submit" from the tool bar. Once the program is submitted, SAS will display messages such as the name of the newly created data set and error or warning messages in the Log window. If the program generates output, it will be displayed in the Output window. You can use the Result window to view and manage different output that is created from your program. To manage files that are stored in the SAS library, you can utilize the Explorer window.

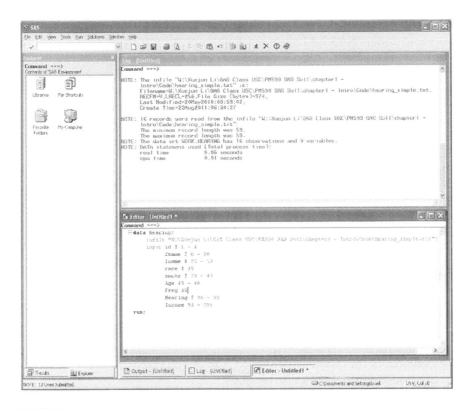

FIGURE 0.1
SAS windowing environment.

SAS Help and Documentation

One of the best resources for learning the SAS language and searching for SAS programming help is *SAS Help and Documentation*, which is part of your SAS software installation. To open *SAS Help and Documentation*, click on the "Help" icon on the tool bar, which will bring up the documents.

There are two windows within *SAS Help and Documentation*. The left window, which contains four tabs (Contents, Index, Search, and Favorites), can help you navigate the different topics. The contents of the selected topic are displayed on the right side of the window. The SAS documentations are grouped by categories under the Contents tab. The Contents tab can often be more useful compared to the other tabs if you know how each topic is categorized.

Most of the references in this book are based on *SAS Help and Documentation*. Every topic within this book is under the "SAS Products ◊ Base SAS" main branch. Within this main branch, if you are looking for documents for non-statistical procedures, it will be under "Base SAS 9.2 Procedure Guide." The topic under "SAS 9.2 Language Reference Concepts" provides detailed documentation about SAS system concepts, SAS DATA steps, and file concepts. Documents that are related to the language elements, such as SAS data set options, formats, functions, informats, statements, and system options, can be found under "SAS 9.2 Language Reference: Dictionary ◊ Dictionary of Language Elements." This search path for SAS documentation through the HELP menu was introduced with SAS Version 9.2. If you are using another version of SAS, such as 9.3, the documentations might be organized differently. Thus, instead of listing the entire search path for a specific documentation, only the name of the last node, such as "Dictionary of Language Elements," is provided as a reference in the book. If you cannot locate the referenced documentation from the HELP menu, an alternative and effective method is to perform a Google search on the name of the documentation (just make sure to include "SAS" in the search string).

How Best to Navigate This Book

The contents of this book are grouped into ten chapters. These ten chapters can be grouped into three main sections.

The first section of the book includes Chapter 1 and Chapter 2, which serve as prerequisite reading for the main section of the book. Chapter 1 provides an introduction and overview of the SAS language, which includes reading data into the SAS system, some Base SAS procedures, creating variables, and subsetting data sets. Some of the topics in these chapters will also be expanded in detail in the later chapters. Chapter 2 discusses how to conditionally create variables based on existing variables, such as how best to use the IF-THEN/ELSE statement or the SELECT statement to create variables.

The second section of the book includes Chapters 3 through 6, which is also the core component of the entire book. These four chapters describe how essential it is to understand the PDV in order to write an accurate program in the DATA step. The topic covered in each chapter is built upon the concepts covered in a previous chapter. Chapter 3 provides an overview about DATA step processing, how to retain variables by using the RETAIN and SUM statements, conditional processing in the DATA step, and how to use the DATA step debugger. Chapter 4 covers BY-group processing within the DATA step and some applications relating to longitudinal data sets. Chapter 5 introduces topics on explicit loops in the DATA step and compares the differences between implicit and explicit loops. Chapter 6 covers array processing, which includes one- to multi-dimensional arrays. Many applications that utilize array processing are related to understanding the explicit loop.

The final section of the book includes Chapters 7 to 10. These four chapters are not built in sequence and can be read independently. Chapter 7 covers multiple methods for combining data sets vertically and horizontally. Although some examples in this chapter demonstrate an understanding of the PDV, readers should be able to grasp and understand most of the materials with or without having read the core section of the book. Chapter 8 covers data input and output. This chapter covers different input methods of reading and writing text files in the DATA step. Chapter 9 covers DATA step functions and call routines, which represent one of the many strengths of SAS software. This chapter introduces a few categories of SAS functions or call routines, such as date and time functions, character functions, and functions for converting variable types. Chapter 10 covers some useful SAS procedures that relate to data management, which include the SORT, COMPARE, TRANSPOSE, FORMAT, and OPTIONS procedures.

A novice programmer with minimum SAS background should read this book cover to cover in the order in which the book is presented. An intermediate SAS programmer who is interested in learning the concept of DATA step processing can begin reading Chapters 3 to 6 without reading the first two prerequisite chapters.

A Note on SAS Output and Log Font

The output and logs that are generated from SAS procedures use the SAS Monospace font. However, the font used for SAS output and logs in this book is Lucida Sans Typewriter. The drawback of using the Lucida Sans Typewriter font is that grid lines for the table displays, such as the output from PROC FREQ, do not have the grid line "look." Here's an example of output from the PROC FREQ using the Lucida Sans Typewriter font:

Output using Lucida Sans Typewriter:

```
                         The FREQ Procedure
                                  Cumulative    Cumulative
   Preg     Frequency      Percent      Frequency       Percent
   ƒƒƒƒƒƒƒƒƒƒƒƒƒƒƒƒƒƒƒƒƒƒƒƒƒƒƒƒƒƒƒƒƒƒƒƒƒƒƒƒƒƒƒƒƒƒƒƒƒƒƒƒƒƒƒƒƒƒƒƒƒ
      0           19        63.33           19         63.33
      1           11        36.67           30        100.00
                     Frequency Missing = 4
```

The "*ƒ*" symbols in the output above appear to be a straight line in the SAS Monospace font. In this situation, the "*ƒ*" symbol is replaced with a dash ("-") to achieve an easier-to-comprehend display of the output, like the one below:

Output using Lucida Sans Typewriter *with modification:*

```
                          The FREQ Procedure
                                     Cumulative    Cumulative
    Preg      Frequency      Percent   Frequency     Percent
    - - - - - - - - - - - - - - - - - - - - - - - - - - - -
     0              19        63.33          19       63.33
     1              11        36.67          30      100.00
                       Frequency Missing = 4
```

Additional Components of This Book

Here is a list of materials that readers can download from the publisher's Web site (http://www.crcpress.com/product/isbn/9781466552388):

- All the data sets used in the book, as well as all the programs
- Exercise data sets and their solutions
- Slides for demonstrating the contents of the PDV for some examples in Chapters 3 through 6

The Author

After receiving an MS in Biostatistics from the University of Southern California (USC), **Arthur Li** embarked on a career as a biostatistician at the City of Hope National Medical Center, in Los Angeles County, California. Li is also a part-time statistical programming instructor at USC. He has given numerous presentations and seminars on DATA step programming and statistical analysis using SAS software at SAS conferences throughout the United States.

Acknowledgments

This book would not have been possible without the tremendous support of my mentor, Prof. Stanley Azen, who provided me the opportunity to teach my SAS class at the Department of Preventive Medicine at the University of Southern California (USC) for the past five years. The accumulated course materials bear an imprint upon this book. I would also like to express my sincere gratitude to Prof. Jim Gauderman, my first SAS teacher, who taught me much while I was his student and his teaching assistant. I also have to thank Prof. Roberta Mckean-Cowdin, who enriched my hands-on SAS experience from the many research projects that were provided by her.

I would also like to thank Xiaolong Li, Han Tun, and Wuchen Zhao for reviewing the contents and testing all the programs in this book.

I would like to acknowledge the enthusiastic and delightful people from the SAS community who provided invaluable support throughout the years for my recognition in the SAS community: MaryAnne DePesquo, Perry Watts, Sunil Gupta, Ron Cody, Kirk Lafler, Peter Eberhardt, and my dear friend Nate Derby. Among them, Sunil Gupta, Ron Cody, and Peter Eberhardt provided many helpful technical suggestions for this book. I am especially deeply indebted to Perry Watts, who spent enormous chunks of her personal time to provide detailed critiques from cover to cover regarding organization, structure, references, and programs/exercises, resulting in a much more precise product.

I am also grateful for the help I received from Taylor & Francis: Rob Calver (Senior Acquisitions Editor), Rachel Holt (Senior Editorial Assistant), Kathryn Everett (Project Coordinator), and Amy Rodriguez (Production Editor).

I would lastly like to express my deepest thanks and gratitude to my beloved Zaccagnino family, especially my "crazy" but lovable in-laws, Dan and Sally Zaccagnino. The Zaccagnino family's endless support makes my success possible. Finally, this book could not have been completed without my partner, David Zaccagnino, who not only edited the ESL (English as a second language) errors in this book but most importantly provided me with endless patience and encouragement.

1

Introduction to SAS®

1.1 SAS Program and Language

A SAS program can consist of one or more DATA steps and procedures (PROC steps), which can be in any sequence depending upon the programming purposes. The building blocks of the DATA and PROC steps are *statements* and are case insensitive. A statement is made up from one or a series of elements, such as SAS names, DATA step functions, operators, operands, special characters, and/or SAS keywords. Each statement must end with a semicolon. As a matter of fact, a missing semicolon in a statement is one of the most common programming errors for novice programmers. Some statements can only be used in either DATA or PROC steps, while some other statements, such as BY, WHERE, LABEL, and FORMAT statements, can be used in both steps. Furthermore, some statements can also be placed outside the DATA or PROC steps; these statements are referred to as being global in scope.

SAS statements that exist in the DATA step mainly serve as an instruction to read external data, create a new SAS data set, or perform data manipulation. Statements that are in the PROC step are mainly used to analyze data or create reports. Global statements are often used to provide information to the SAS system, such as modifying the appearance of the SAS log, adding titles or footnotes to the output, or controlling the methods as to how SAS processes your program.

In addition to statements, there are other types of SAS language elements, such as data set options, expressions, formats, and informats. These language elements are mostly used within a statement and will be introduced when they are used within a certain SAS statement throughout this book.

In order to be consistent in your SAS documentation, the syntax convention in this book follows the same convention provided in the *SAS® 9.2 Language Reference: Dictionary* (2009). The description of the syntax convention is described in the "Syntax Conventions for the SAS Language" article in SAS documentation.

1.2 Reading Data into SAS

The starting point for most projects that a programmer becomes involved with is reading data into the SAS system. The most frequently used input file formats a programmer will encounter are SAS data sets that already exist, raw text files, and Microsoft® EXCEL spreadsheets.

A SAS data set is often created by reading, extracting, or combining data from text files, EXCEL files, or some other SAS data sets. Once a SAS data set is created, programmers tend to save it as a permanent file. Not only is re-reading a SAS data set simple, but the SAS data set is also easier to manage than other types of file formats. External text files or EXCEL spreadsheets, for example, can be confusing if not enough attention is paid to how the data are arranged on a page. This section covers reading a SAS data set and reading a raw data file with fixed fields. Reading raw data with other types of formats will be covered in Chapter 8.

1.2.1 The SAS Data Set and SAS Library

A SAS data set is a SAS file that is created by SAS software. A SAS data set can be either a SAS data file or a SAS view. A SAS data view, which is not covered in this book, is a virtual data set pointing to the data from other sources. A SAS data file contains data and the descriptor information of the data. In this book, the terminologies for *data set* and *data file* are used interchangeably.

The values stored in the SAS data set are arranged in a table of rows and columns. Rows are referred to as observations or records, and columns are referred to as variables. In SAS, there are two types of variables: character and numeric. Character variables can contain alphanumeric values or special characters. The missing values for character values are presented as blanks. Numeric variables can contain floating-point numbers. The missing values for numeric values are represented as periods (.). In SAS, variables that contain date and time values are also considered numeric variables.

The descriptor information of a data set includes information about the creation date of the data set, the number of observations, and the attributes of each variable, such as a variable's name, length, type, label, format, informat, etc.

All files, including the SAS data sets on your computer, are stored and organized in different directories and subdirectories. In SAS, the directory that stores SAS data sets is referred to as a *library*. To access or create a SAS file in a library, you need to begin with the LIBNAME statement, which has the following form:

```
LIBNAME libref 'SAS-library';
```

The LIBNAME statement starts with the keyword LIBNAME and is followed by *libref* and '*SAS-library*'. *Libref* is the logical library name, while '*SAS-library*' is the physical location of the file folder. The basic idea of using the LIBNAME statement is associating a logical SAS library name, *libref*, with the directory path in which permanent SAS data sets are stored. Thus, in your SAS program, instead of referring to the file's directory with the complete path name, you will use *libref*. For example, in the following SAS statement, the *libref* is SASLIB, which is associated with the physical location 'C:\SAS Book\dat':

```
libname saslib 'C:\SAS Book\dat';
```

LIBNAME is a global statement and is used outside the DATA and PROC steps. Being *global* in this situation means that the name of the library, *libref*, is only in effect until you change it, cancel it, or terminate your current SAS session. The contents of the library always exist unless you delete them.

To access the data set in a library, you need to use a two-level name, which has the following form:

```
libref.filename
```

The first component of the two-level name is the library name, and the second component is the name of the SAS data set; the two names are separated by a period (.). For example, SASLIB.MYDAT refers to the SAS data set MYDAT that is stored in the SASLIB library.

The rules for naming *libref* and *filename* are similar. The starting position of both names must begin with either an underscore or a letter. The nonstarting position of the names can contain an underscore, letters, and numbers. The length of the *libref* cannot be more than eight characters, and the length of the *filename* can be up to 32 characters.

You can also use a single-level name by providing *filename* only and omitting *libref*. In this situation, you are accessing or creating a SAS data set stored in the WORK library. The WORK library is automatically created when you start your SAS session. The WORK library is a temporary library that is used to store temporary SAS files. What this means is that all files stored in the WORK library are deleted when the current session terminates. You can refer to the data set that is stored in the WORK library by using either a single-level name (*filename*) or a two-level name (WORK.*filename*). However, to refer to a data set that is stored in the permanent library, you must always use a two-level name (*libref.filename*).

1.2.2 Reading a SAS Data Set

Regardless of your programming purpose, writing a SAS program utilizing a DATA step will create one or more SAS data sets. For example, when you are reading a SAS data set into SAS, you are creating a SAS data set that is based on the SAS data set that you are reading.

A DATA step always starts with the DATA statement and usually ends with a RUN statement. To read a SAS data set into SAS, you need to use at least the following three statements:

```
DATA output-data-set-name;
   SET input-data-set-name;
RUN;
```

In the DATA statement, *output-data-set-name* is the name of the data set that you are creating, and the *input-data-set-name* in the SET statement is the name of the data set that you are reading. Both *output-data-set-name* and *input-data-set-name* can be either a one-level or two-level name. In this book, the data set being created after the keyword DATA is called the *output data set*, and the data set you are reading after the keyword SET is called the *input data set*. The RUN statement is used to execute the previously entered statements.

Notice that the SET statement is indented. SAS is an indentation-insensitive language. Indenting only serves to add visual clarity to your code. Other than the first and last statements of a DATA or a PROC step, programmers often indent the remainder of the statement so that each DATA or a PROC step can be distinguished individually.

To further clarify your program, you can provide documentation to your program by using the COMMENT statement. The COMMENT statement takes one of the following forms:

```
*message;
/*message*/
```

The *message* component in both formats is used to specify the text that explains the statement or program. SAS ignores text in COMMENT statements during processing. The *message* in the first format is enclosed between an asterisk (*) and a semicolon (;), and the *message* in the second format is enclosed between a slash star (/*) and star slash (*/). The main difference between these two formats is that the first one, starting with an asterisk and ending with a semicolon, cannot contain internal semicolons or unmatched quotations, while the second format can. Furthermore, you should avoid placing /* comment symbols in columns 1 and 2 of your code, because in some operating environments, SAS might interpret a /* in columns 1 and 2 as a request to terminate your SAS program.

Program 1.1 creates two SAS data sets by reading NOISE.SAS7BDAT from the SASLIB library. All the SAS data sets end with ".SAS7BDAT". For purposes of simplicity, the SAS file extension will be omitted for the remainder of this book. In Program 1.1, the first DATA step creates a temporary data set (NOISE1) that is stored in the library WORK. The second DATA step creates a permanent data set (NOISE1) that is stored in the DESKTOP library. Notice that both data sets being created have the same data set names but are stored in different libraries. This program also utilizes two types of COMMENT statements that provide further explanation of the program.

Program 1.1:

```
libname saslib 'W:\SAS Book\dat';
libname desktop 'C:\Documents and Settings\Desktop';
data noise1; *creates NOISE1.SAS7BDAT in the library WORK;
    set saslib.noise;
run;


data desktop.noise1;/*becomes NOISE1.SAS7BDAT in w:\SAS Book\
dat */
    set saslib.noise;
run;
```

The purpose of Program 1.1 is to provide more illustrations, because you would never actually write such a program since you are simply making a copy of the data set. If you would like to test this program, you need to download the data set NOISE on your computer and specify the correct directory path in which NOISE is located on your computer in the LIBNAME statement. The information regarding downloading all the testing data sets and the program is described in the Preface.

Once a program is submitted, the first thing that you should check is the SAS log that is generated from the submitted SAS code. The SAS log often contains information about submitted programs, including warning or error messages. For example, the SAS Log from Program 1.1 contains the message about the number of observations being read from the NOISE data set, the number of observations and variables in the output data sets, and DATA step computing times.

Log from Program 1.1:

```
13 libname saslib 'W:\SAS Book\dat';
NOTE: Libref SASLIB was successfully assigned as follows:
      Engine:       V9
      Physical Name: W:\SAS Book\dat
14 libname desktop 'C:\Documents and Settings\Desktop';
NOTE: Libref DESKTOP was successfully assigned as follows:
      Engine:       V9
      Physical Name: C:\Documents and Settings\Desktop
15 data noise1; *creates NOISE1.SAS7BDAT in the library WORK;
16     set saslib.noise;
17 run;


NOTE: There were 32 observations read from the data set
SASLIB.NOISE.
NOTE: The data set WORK.NOISE1 has 32 observations and 4
variables.
```

```
NOTE: DATA statement used (Total process time):
      real time            0.04 seconds
      cpu time             0.01 seconds

18 data desktop.noise1;/*becomes NOISE1.SAS7BDAT in W:\SAS
Book\dat
19 ! */
20    set saslib.noise;
21 run;

NOTE: There were 32 observations read from the data set
SASLIB.NOISE.
NOTE: The data set DESKTOP.NOISE1 has 32 observations and 4
variables.
NOTE: DATA statement used (Total process time):
      real time            0.00 seconds
      cpu time             0.00 seconds
```

1.2.3 Reading a Raw Data File with Fixed Fields

In this book, a raw data file is also referred to as a text file. A raw data file with fixed fields means that the values of the variables occupy the same location for all the observations. For example, HEARING.TXT contains variables in the fixed fields (only the first five observations are shown). The first row contains the column numbers; they are not part of the data file. The description of each variable, along with their column information, is shown in Table 1.1.

HEARING.TXT:

```
12345678901234567890123456789 0
629F H past     26 0        35000
656F W never    26 1 no     48000
711F W never    32 1 no     30000
733F W current 17 0         59000
135F B current 29 1 no     120000
```

TABLE 1.1

Variable Information for HEARING.TXT

Variable Name	Description	Locations	Variable Type
ID	Subject's four-digit ID	Columns 1–4	Character
RACE	Ethnicity	Column 6	Character
SMOKE	Smoking status	Columns 8–14	Character
AGE	Age	Columns 16–17	Numeric
PREG	Pregnancy	Column 19	Numeric
HEARING	Hearing loss	Columns 21–23	Character
INCOME	Annual income	Columns 25–30	Numeric

Reading an external text file often starts with the INFILE statement, which is used to identify the location of the text file. The INFILE statement has the following form:

```
INFILE file-specification <OBS=record-number>;
```

File-specification is used to specify the location of the input data; the common form is the physical file name along with its full path. The language elements that are enclosed between a less than sign (<) and a greater than sign (>) are *options* for the SAS statement. In this example, the optional OBS option is used to specify the first number of records to be read. This option is especially useful for reading a data set with a large number of observations. For example, you can read the first few observations to verify if the data is being read correctly before continuing to read the entire data set.

Once the file location is identified from the INFILE statement, you need to use the INPUT statement to read the external text file. In SAS, there are four types of input methods: *column, formatted, list,* and *named.* In this section, only the column input method is presented. The expansion of this topic, along with other input methods, will be covered in Chapter 8.

You can use the column input method to read raw data that contains variables in a fixed field with character or standard numeric values. The standard numeric data values include numbers, decimal points, numbers in scientific notations (e.g., 1.2E4), and plus or minus signs. Nonstandard numeric data can include date and time values, fractions, integers, real binary numbers in hexadecimal forms, and values that contain special characters, such as %, $, and comma (,). Here is the syntax for the column input method:

```
INPUT variable <$> start-column <- end-column>
```

In the INPUT statement, *variable* is the name of the variable that you are creating by associating it with the input values from specified columns. The naming convention for the *variable* is the same as the rules for naming a SAS data set. The optional dollar sign ($) is used when creating a character variable from character values. *Start-column* and *end-column* are the starting and ending positions of the input values. *End-column* is optional if the variable value occupies only one field. Program 1.2 reads the HEARING.TXT file by using the column input method.

Program 1.2:

```
data hearing;
    infile "W:\SAS Book\dat\hearing.txt";
    input id $ 1 - 4
          race $ 6
          smoke $ 8 - 14
          Age 16 - 17
```

```
            Preg 19
            Hearing $ 21 - 23
            Income 25 - 30;
run;
```

1.2.4 Reading Data Entered Directly into the Program

For a data set with a small number of observations and variables, you can enter the data directly into the DATA step by using the DATALINES statement, which has the following form:

DATALINES;

When reading data directly from the DATA step, you need to place the DATALINES statement in the last statement of the DATA step and enter your data immediately after the DATALINES statement. In the end, you need to write a single semicolon (a NULL statement) to indicate the end of the input data. If the data that you entered contains semicolons, you must use the DATALINES4 statement and use four consecutive semicolons (;;;;) instead of one at the end of the input data. Program 1.3 illustrates the use of the DATALINES statement by reading only the first 16 observations from the HEARING data set.

Program 1.3:

```
data hearing_small;
    input id $ 1 - 4
            race $ 6
            smoke $ 8 - 14
            Age 16 - 17
            Preg 19
            Hearing $ 21 - 23
            Income 25 - 30;
datalines;
629F H past     26 0         35000
656F W never    26 1 no      48000
711F W never    32 1 no      30000
733F W current  17 0         59000
135F B current  29 1 no     120000
982F W past     26 1 yes    113000
798F W never    19 0         28900
494F W never    36 0         65000
748F W never    34 1 no      39000
904F W never    25 0         76200
244F W never    28 1 yes     58000
747F A current  18 0         39000
796F A past     35 0        134000
713F H never    26 1 no      29000
```

```
745F A never   36 1 no    76000
184M W past    19 .       13900
;
```

1.3 Creating and Modifying Variables

After data is read into the SAS system, an immediate common task is to create or modify variables because the existing variables from the input data do not always contain the information that you need. This section covers creating variables by using the *assignment statement* and introduces how to create variables conditionally. Conditionally creating variables will be further discussed in Chapter 2.

1.3.1 The Assignment Statement and SAS Expression

The *assignment* statement is the most common method used for creating a variable. It has the following form:

```
variable=expression;
```

In the assignment statement, *variable* is either a new or existing variable, and *expression* is any valid SAS expression.

The purpose of using an expression in a SAS statement is to create variables, assign values, perform calculations, transform variables, and perform conditional processing. All expressions will return a result with a character, numeric, or Boolean value. An expression is formed by a sequence of *operands* and *operators*. Operands are either constants or variables. Operators include symbols for arithmetic calculations, comparisons, logical operations, SAS functions, or grouping parentheses.

Examples of operators, along with their evaluation orders, are illustrated in Table 1.2. Operations at priority 1 level are executed before priority 2 level, and priority 2 level operations are executed before priority 3 level, and so on. The order of operations can also be controlled by parentheses. You should always use parentheses when you are not sure about the operation order. Some operators have mnemonic-equivalent forms that provide alternative forms that you can use instead of the corresponding operators.

Comparison operators are often used with IF-THEN/ELSE statements. The use of comparison operators is for comparison or calculation purposes between two variables, constants, or expressions. If the comparison is true, the result is returned with 1; otherwise, the result is returned with 0. The IN operator is used to determine whether a variable's value is among the list of character or numeric values. When character values are used for comparison,

TABLE 1.2

Definition, Evaluation Order, and Examples of Operators

Priority	Evaluation Order	Type	Operator	Mnemonic Equivalent	Definition	Example
1	Right to left	Arithmetic	−		Negation prefix	`b = -a;`
		Arithmetic	**		Exponentiation	`b2 = a**2;`
		Logical	^ ~[a]	NOT	Logical not	`~z`
2	Left to right	Arithmetic	*		Multiplication	`c = a*b;`
		Arithmetic	/		Division	`c = a/b;`
3	Left to right	Arithmetic	+		Addition	`c = a+b;`
		Arithmetic	−		Subtraction	`c = a-b;`
4	Left to right	Comparison	<	LT	Less than	`a<10`
		Comparison	<=	LE	Less than or equal to	`a le b`
		Comparison	=	EQ	Equal to	`b = 2`
		Comparison	^=	NE	Not equal to	`z ne 'A'`
		Comparison	>=	GE	Greater than or equal to	`g> = c`
		Comparison	>	GT	Greater than	`g gt a`
				IN	equal to one of a list	`z in ('A', 'B', 'G')`
5	Left to right	Logical	&	AND	Logical and	`a<10 & a>5`
6	Left to right	Logical	|[b]	OR	Logical or	`z = 'A'|z = 'B'`

[a] Use either ^ or ~ for a "logical not" operator.
[b] Use either \| or ! for a "logical or" operator.

they must be written in the same case as they appear in the original data set and must be enclosed in either single or double quotation marks.

An expression can be categorized into *simple, compound,* and *WHERE* (discussed in Section 1.4.1). A simple expression can contain no more than one operator, while a compound expression can contain more than one. You can connect one or more simple expressions with logical operators to form a compound expression.

A SAS function can also be considered an operator because a function performs a certain type of calculation and returns a value. More details about SAS functions are discussed in Chapter 9. You can use SAS functions in DATA step statements or in a WHERE expression. Here's the general form of a SAS function:

```
function-name(argument-1<,...argument-n>)
```

For example, to calculate the sum of a list of numeric variables, you can use the SUM function, which has the following form:

```
SUM(argument-1<,...argument-n>)
```

Program 1.4 reads five observations by using the DATALINES statement and creates two variables: SCORE_SUM1 and SCORE_SUM2. SCORE_SUM1 and SCORE_SUM2 are created by using the addition (+) operator and the SUM function, respectively. When adding two or more variables, the SUM function treats the missing values as zero; hence, the resulting value will not be missing. However, calculations that result by the use of the arithmetic operator will result in a missing value if any operand contains a missing value for an observation.

Program 1.4:

```
data score;
    input ID $ 1-4 score1 6-7 score2 9-10 score3 12-13;
    score_sum1 = score1 + score2 + score3;
    score_sum2 = sum(score1, score2, score3);
datalines;
629F 5  6  9
656F 6 10  9
711F 0  .  3
511F 9  4 10
478F .  5  3
;

title 'Adding Three Scores By Using + Operator and SUM
Function';
proc print data = score;
run;
```

Output from Program 1.4:

```
Adding Three Noise Scores By Using + Operator and SUM Function
Obs    ID     noise1    noise2    noise3    noise_ sum1    noise_ sum2
1     629F       5         6         9           20            20
2     656F       6        10         9           25            25
3     711F       0         .         3            .             3
4     511F       9         4        10           23            23
5     478F       .         5         3            .             8
```

The PRINT procedure immediately following the DATA step is used to print the contents of the data set. The TITLE statement is used to add the title for the output generated from using PROC PRINT. Further details concerning the use of PROC PRINT and the TITLE statement will be presented in Section 1.4.

1.3.2 Creating Variables Conditionally

In many situations, you need to create variables conditionally, which can be done by using the IF-THEN/ELSE statement. This topic is expanded upon in Chapter 2. The IF-THEN/ELSE statement has the following form:

```
IF expression THEN statement;
<ELSE statement;>
```

The *expression* in the IF-THEN/ELSE statement can be any valid SAS expression and often contains a comparison operator. In the situation where you need to have more than one expression to form a condition, you can use logical operators to create compound expressions. If the *expression* is evaluated to be true, the IF-THEN statement executes the *statement* after the keyword THEN for observations that are read from a data set. If there is an optional ELSE statement and if the *expression* is evaluated to be true, the ELSE statement is not executed; otherwise, the ELSE statement is executed. To use the ELSE statement, place it immediately after the IF-THEN statement.

Suppose that you would like to create a variable, named OVER10K. If income is greater then 10,000, OVER10K will be assigned to 1; otherwise OVER10K will be assigned to 0. There are multiple ways of creating this variable, for example:

```
if income > 10000 then over10k = 1;
if income < = 10000 then over10k = 0;
```

These two statements are legitimate; however, the second statement is executed even if the condition in the first statement is evaluated to be true. A more efficient way to create these variables is by writing the following code:

```
if income > 10000 then over10k = 1;
else over10k = 0;
```

By adding the ELSE statement, if the condition in the IF statement is evaluated to be true, the second statement will not be processed.

In SAS, a missing value is the smallest value. Thus, in the example above, if the INCOME variable contains any missing values, the observations with the missing values will be assigned to 0 for the OVER10K variable. How best to handle missing values when creating variables is discussed in Section 2.1.2.

1.4 Base SAS Procedures

Base SAS software provides a large selection of procedures that allow you to examine the contents of a SAS data set. Detailed documentation can be found in *Base SAS® 9.2 Procedures Guide* (2009). These procedures can be grouped into three categories: report writing, statistics, and utilities. This chapter covers only a few commonly used procedures: CONTENTS, SORT, PRINT, MEANS, and FREQ. The procedures in this section will be frequently used throughout the book as data-checking tools.

1.4.1 Common Statements in SAS Procedures: TITLE, BY, and WHERE Statements

This section covers only the two most commonly used statements, the BY and WHERE statements, to reduce repeated or redundant future explanations when introducing other procedures that use these statements. In addition, the TITLE statement, which is a global statement, is discussed in this section because it is frequently used when generating output.

Most SAS procedures will generate output in the SAS output window after procedures are submitted. By default, the generated output will have "The SAS System" as its title. To override the default title, you can use the TITLE statement, which has the following form:

```
TITLE <'text' | "text">;
```

The optional *text* is the new title that you may want to display for the procedure, which can be in either single or double quotations. You can place the TITLE statement within or outside the procedure. Once a title statement is submitted, it will be used for all subsequent output until you cancel or create a new title. To cancel a title, you can simply submit a TITLE statement without adding any text.

Many SAS statements have the same functionality across procedures, such as the BY and WHERE statements. Using the BY statement will categorize the output by each level of the variable that is used in the BY statement. The BY statement has the following form:

```
BY <DESCENDING> variable-1 <... <DESCENDING> variable-n>;
```

You can list more than one variable in a BY statement. The *variable* specified in the BY statement is called a *BY variable*. If the BY statement is used in a procedure, the output will be grouped by each level of the variable(s) used in the BY statement. The DESCENDING option is used to sort the *variable* that immediately follows the DESCENDING keyword in descending order. Here are some of the procedures that support the BY statement: REPORT, SORT (required), COMPARE, CORR, FREQ, TABULATE, MEANS, PLOT, TRANSPOSE, PRINT, UNIVARIATE, etc.

The WHERE statement subsets the input data set by specifying conditions a record has to meet for inclusion. Only those observations that meet the specified conditions will be used for processing. The WHERE statement has the following form:

```
WHERE where-expression;
```

The *where-expression* needs to be a legitimate arithmetic or logical expression. Here are some of the procedures that support the WHERE statement: REPORT, COMPARE, SORT, CORR, FREQ, TABULATE, MEANS, PLOT, TRANSPOSE, PRINT, UNIVARIATE, etc.

1.4.2 The CONTENTS Procedure

All SAS data sets contain a descriptor portion that contains information about the data set. PROC CONTENTS not only displays the descriptor information of a specified data set, it can also show the contents of a specified SAS library. The basic syntax for PROC CONTENTS is as follows:

```
PROC CONTENTS <DATA=SAS-file-specification> <VARNUM>;
RUN;
```

All SAS procedures contain a DATA=option, which is used to specify the name of the input data set for the procedure. Without using the DATA = option, SAS will use the most recently created data set. The *SAS-file-specification* can be in one of the following forms:

```
<libref.>SAS-data-set
<libref.>_ALL_
```

When *<libref.>SAS-data-set* is used, PROC CONTENTS displays the contents of *SAS-data-set* within the given library. Without specifying *libref* explicitly, *libref* is referred to as the WORK library. Using *<libref.>_ALL_*, PROC

CONTENTS will list all SAS data sets along with their information specified by the given library.

The VARNUM option prints the variable names by their created order. By default, variable names are listed alphabetically. Program 1.5 uses PROC CONTENTS to display the descriptor portion of the HEARING data set.

Program 1.5:

```
title 'The Contents of Hearing Data';
proc contents data = hearing varnum;
run;
```

Output from Program 1.5:

```
                  The Contents of Hearing Data
                     The CONTENTS Procedure

Data Set Name         WORK.HEARING        Observations           34
Member Type           DATA                Variables              7
Engine                V9                  Indexes                0
Created               Sunday, January     Observation Length     40
                      29, 2012 09:36:46 AM
Last Modified         Sunday, January     Deleted Observations   0
                      29, 2012 09:36:46 AM
Protection                                Compressed             NO
Data Set Type                             Sorted                 NO
Label
Data Representation   WINDOWS_32
Encoding              wlatin1 Western
                      (Windows)

                 Engine/Host Dependent Information
Data Set Page Size            4096
Number of Data Set Pages      1
First Data Page               1
Max Obs per Page              101
Obs in First Data Page        34
Number of Data Set Repairs    0
Filename                      C:\Users\Arthur\AppData\Local\
                              Temp\SAS Temporary
                              Files\_TD3988\hearing.sas7bdat
Release Created               9.0202M0
Host Created                  W32_VSPRO

                 Variables in Creation Order
               #    Variable   Type     Len
               1    id         Char      4
               2    race       Char      1
               3    smoke      Char      7
```

4	Age	Num	8
5	Preg	Num	8
6	Hearing	Char	3
7	Income	Num	8

1.4.3 The SORT Procedure

The SORT procedure is used to order the data set observations by the values of one or more variables. In SAS, a missing value, either numeric or character, is the smallest value. The general syntax for PROC SORT is as follows:

```
PROC SORT <DATA=SAS-data-set> <OUT=SAS-data-set>;
    BY <DESCENDING> variable-1 <... <DESCENDING> variable-n>;
RUN;
```

In the PROC SORT statement, the DATA=*SAS-data-set* option is used to identify the input SAS data set for sorting, and the OUT=*SAS-data-set* option is used to name the output data set that contains newly sorted data. If the OUT= option is omitted, the PROC SORT procedure will replace the original data with the sorted data.

At least one variable is required in the BY statement. By default, the SORT procedure orders the data set by the values of the first BY variable in ascending order. If there is more than one BY variable, the SORT procedure orders the observations having the same value of the first BY variable by the values of the second BY variable in ascending order. The sorting scheme will continue for all the BY variables. The DESCENDING option sorts the variable that follows the DESCENDING keyword in descending order. For example, Program 1.6 sorts RACE in ascending order, PREG in descending order, and AGE in descending order.

Program 1.6:

```
proc sort data = hearing out = hearing_sort;
    by race descending preg descending age;
run;
```

1.4.4 The PRINT Procedure

The PRINT procedure is one of the report-writing procedures that you can utilize to generate a customized report. Only the basic usage of PROC PRINT is presented in this section. More often, you will use PROC PRINT to simply print the observations for a specified SAS data set using either all or some of the variables. The basic syntax for PROC PRINT is as follows:

```
PROC PRINT <DATA=SAS-data-set> <NOOBS>;
    BY <DESCENDING> variable-1 <... <DESCENDING> variable-n>;
```

```
   VAR variable(s);
   WHERE where-expression;
RUN;
```

By default, SAS prints the observation numbers in the output. Using the NOOBS option will suppress the observation numbers.

If the optional BY statement is used, PROC PRINT will create a separate section of the report for each level of the variables specified in the BY statement. Furthermore, the data set needs to be previously sorted (by the same order) as listed in the BY statement. You can use multiple variables in the BY statement.

The optional VAR statement is used to select variables to be printed in the output. If the VAR statement is omitted, all the variables will be printed in the output. The optional WHERE statement is used to subset the input data set by specifying certain conditions.

For example, Program 1.7 prints the contents of the HEARING_SMALL data set by each category of the RACE variable.

Program 1.7:

```
proc sort data = hearing_small out = hearing_small_sort;
    by race;
run;

title 'Print ID SMOKE and AGE variables by RACE';
proc print data = hearing_small_sort noobs;
    by race;
    var id smoke age;
run;
```

Output from Program 1.7:

```
        Print ID SMOKE and AGE variables by RACE

– – – – – – – – – – – – – – race = A– – – – – – – – – – – – –
                id          smoke          Age
                747F        current        18
                796F        past           35
                745F        never          36


– – – – – – – – – – – – – – race = B– – – – – – – – – – – – –
                id          smoke          Age
                135F        current        29


– – – – – – – – – – – – – – race = H– – – – – – – – – – – – –
                id          smoke          Age
                629F        past           26
                713F        never          26
```

```
_ _ _ _ _ _ _ _ _ _ _ _ _ _ race = W- _ _ _ _ _ _ _ _ _ _ _ _ _ _
          id            smoke           Age
          656F          never           26
          711F          never           32
          733F          current         17
          982F          past            26
          798F          never           19
          494F          never           36
          748F          never           34
          904F          never           25
          244F          never           28
          184M          past            19
```

1.4.5 The MEANS Procedure

You can use PROC MEANS to calculate descriptive statistics for numerical variables across all observations in the input data set. PROC MEANS has the following form:

```
PROC MEANS <DATA=SAS-data-set> <MAXDEC=number>
           <statistic-keyword(s)>;
   BY <DESCENDING> variable-1 <... <DESCENDING> variable-n>;
   CLASS variable(s);
   VAR variable(s);
   WHERE where-expression;
RUN;
```

The MAXDEC=*number* option in the PROC MEANS statement is used to specify the maximum number of decimal places to display statistics in the output. Without specifying this option, the default decimal place is seven. You can specifically request the statistics that you would like to compute by specifying statistic keyword(s) in the *statistic-keyword(s)* option. The commonly used keywords are NMISS, N, RANGE, STD, MAX, MEAN, MIN, VAR, MEDIAN, Q1, Q3, etc. Without specifying *statistic-keyword(s)*, the default statistics (N, MEAN, STD, MIN, and MAX) will be computed.

Similar to other procedures, the optional BY statement computes separate statistics for each BY group; likewise, the data set needs to be sorted previously in the same order as listed in the BY statement.

The optional CLASS statement is used to specify the variables whose values define the subgroup combinations for the analysis. The use of the CLASS statement is similar to the BY statement; however, the output created by using the CLASS statement has a different layout. You don't need to sort the *variable(s)* specified in the CLASS statement first before using the MEAN procedure.

The optional VAR statement is used to list the variable(s) for computing statistics. Omitting the VAR statement will result in SAS computing statistics for all the numeric variables that are not listed in the other statements.

The optional WHERE statement is used to subset the input data set by specifying certain conditions.

Program 1.8 calculates minimum, median, and maximum values for AGE variables just for the White and Black groups from the HEARING data set. These computed statistics are separated by each level of the SMOKE variable. The first PROC MEANS uses the CLASS statement for the SMOKE variable, and the second PROC MEANS uses the BY statement for the SMOKE variable. The output generated from Program 1.8 shows the differences between using the CLASS and BY statements.

Program 1.8:

```
title 'The Mean Procedure - Class Statement';
proc means data = hearing min median max maxdec = 2;
    where race = 'W' or race = 'B';
    class smoke;
    var age;
run;

proc sort data = hearing;
    by smoke;
run;

title 'The Mean Procedure - By Statement';
proc means data = hearing min median max maxdec = 2;
    where race = 'W' or race = 'B';
    by smoke;
    var age;
run;
```

Output from Program 1.8:

```
                The Mean Procedure - Class Statement
                      The MEANS Procedure
                   Analysis Variable : Age
smoke        N Obs        Minimum        Median        Maximum
- - - - - - - - - - - - - - - - - - - - - - - - - - - - - - - - -

current        6           17.00          28.50          33.00
never         14           19.00          27.00          36.00
past           5           19.00          21.00          34.00
- - - - - - - - - - - - - - - - - - - - - - - - - - - - - - - - -


                The Mean Procedure - By Statement
- - - - - - - - - - - - - smoke = current- - - - - - - - - - -
                      The MEANS Procedure
                   Analysis Variable : Age
        Minimum            Median            Maximum
     - - - - - - - - - - - - - - - - - - - - - - - - -

        17.00              28.50              33.00
     - - - - - - - - - - - - - - - - - - - - - - - - -
```

```
                      Using MISSPRINT option
                       The FREQ Procedure
                                      Cumulative      Cumulative
   Preg      Frequency     Percent     Frequency        Percent
   - - - - - - - - - - - - - - - - - - - - - - - - - - - -

     .            4           .             .               .
     0           19         63.33          19             63.33
     1           11         36.67          30            100.00
                      Frequency Missing = 4
```

Program 1.10 illustrates the difference in the output when using (or not using) the LIST option.

Program 1.10:

```
title 'Without LIST option';
proc freq data = hearing;
    tables preg*race*smoke;
run;

title 'With LIST option';
proc freq data = hearing;
    tables preg*race*smoke/list;
run;
```

Output from Program 1.10:

```
                      Without LIST option
                       The FREQ Procedure
                     Table 1 of race by smoke
                     Controlling for Preg = 0
        race        smoke
        Frequency|
        Percent  |
        Row Pct  ,
        Col Pct  |current |never  |past    | Total
        - - - -  -^- - - -^- - - -^- - - -^
        A        |     2 |     1 |     1 |     4
                 | 10.53 |  5.26 |  5.26 | 21.05
                 | 50.00 | 25.00 | 25.00 |
                 | 33.33 | 11.11 | 25.00 |
        - - - - -+- - - - + - - - + - - - +
        B        |     1 |     2 |     0 |     3
                 |  5.26 | 10.53 |  0.00 | 15.79
                 | 33.33 | 66.67 |  0.00 |
                 | 16.67 | 22.22 |  0.00 |
        - - - - -+- - - - + - - - + - - - +
        H        |     0 |     1 |     1 |     2
                 |  0.00 |  5.26 |  5.26 | 10.53
                 |  0.00 | 50.00 | 50.00 |
                 |  0.00 | 11.11 | 25.00 |
```

```
- - - - -+- - - + - - - + - - - +
W          |        3 |      5 |      2 |    10
           |    15.79 |  26.32 |  10.53 | 52.63
           |    30.00 |  50.00 |  20.00 |
           |    50.00 |  55.56 |  50.00 |
- - - - -+- - - + - - - + - - - +
Total              6         9        4       19
               31.58     47.37    21.05   100.00
```

                       Without LIST option
                        The FREQ Procedure
                      Table 2 of race by smoke
                       Controlling for Preg = 1

```
race       smoke
Frequency|
Percent  |
Row Pct  |
Col Pct  |current |never   |past    | Total
- - - -  -+- - - + - - - -+- - - +
A          |      0 |      1 |      0 |     1
           |   0.00 |  10.00 |   0.00 | 10.00
           |   0.00 | 100.00 |   0.00 |
           |   0.00 |  12.50 |   0.00 |
- - - -  -+- - - + - - - -+- - - +
B          |      1 |      0 |      0 |     1
           |  10.00 |   0.00 |   0.00 | 10.00
           | 100.00 |   0.00 |   0.00 |
           | 100.00 |   0.00 |   0.00 |
- - - -  -+- - - + - - - -+- - - +
H          |      0 |      1 |      0 |     1
           |   0.00 |  10.00 |   0.00 | 10.00
           |   0.00 | 100.00 |   0.00 |
           |   0.00 |  12.50 |   0.00 |
- - - -  -+- - - + - - - -+- - - +
W          |      0 |      6 |      1 |     7
           |   0.00 |  60.00 |  10.00 | 70.00
           |   0.00 |  85.71 |  14.29 |
           |   0.00 |  75.00 | 100.00 |
- - - -  -+- - - +- - - - + - - - +
Total              1        8        1       10
               10.00    80.00    10.00   100.00
                   Frequency Missing = 1
```

                       With LIST option
                        The FREQ Procedure

```
                                      Cumulative Cumulative
Preg  race   smoke   Frequency  Percent   Frequency    Percent
- - - - - - - - - - - - - - - - - - - - - - - - - - - - - -
  0   A     current        2      6.90           2       6.90
  0   A     never          1      3.45           3      10.34
```

0	A	past	1	3.45	4	13.79
0	B	current	1	3.45	5	17.24
0	B	never	2	6.90	7	24.14
0	H	never	1	3.45	8	27.59
0	H	past	1	3.45	9	31.03
0	W	current	3	10.34	12	41.38
0	W	never	5	17.24	17	58.62
0	W	past	2	6.90	19	65.52
1	A	never	1	3.45	20	68.97
1	B	current	1	3.45	21	72.41
1	H	never	1	3.45	22	75.86
1	W	never	6	20.69	28	96.55
1	W	past	1	3.45	29	100.00

Frequency Missing = 5

1.5 Subsetting Data by Selecting Variables

In SAS, *subsetting* by data refers to selecting a portion of a SAS data set by keeping only certain variables or selected observations (or both). This section covers how to subset a data set by selecting certain variables. Subsetting a data set by selecting observations is covered in Chapter 3.

When you subset data by selecting variables, you often need to keep or drop a list of variables. SAS provides a convenient method that allows you to refer to a list of variables by using the SAS *variable-lists* notation. Variable list notation can be used in many SAS statements and data set options. There are four types of variable-lists: *numbered range lists, name range lists, name prefix lists,* and *special SAS name lists.*

The variables that are referenced in *numbered range lists* must have the same name, except for the last one, two, or more characters. The last one or more characters need to be consecutive numbers. You need to use a single dash to connect the beginning and ending variables. For example, writing VAR1 - VAR3 is equivalent to writing VAR1, VAR2, VAR3 or VAR1 VAR2 VAR3.

To use *named range lists,* you need to know the creation order of the data set, which can be found via the VARNUM option from PROC CONTENTS. For example, ID -- PREG refers to all the variables in order of variable creation from ID to PREG.

Use *name prefix lists* to refer to all the variables that begin with a specified character string. For example, DRUG: refers to all the variables that begin with "DRUG."

Last, *special SAS name lists* uses special SAS names, _NUMERIC_, _CHARACTER_, and _ALL_, to refer to all numeric, character, or any variables that are already defined in the current DATA step, respectively. Examples of using *variable-lists* are summarized in Table 1.4.

TABLE 1.4

Examples of SAS Variable-Lists

Variable List	Example	Equivalence
Numbered range lists	VAR1 - VAR3	VAR1, VAR2, VAR3 or VAR1 VAR2 VAR3
Named range lists	ID -- PREG	All the variables in order of variable creation from ID to PREG
	ID -NUMERIC- PREG	All numeric variables from ID to PREG
	ID -CHARACTER- PREG	All character variables from ID to PREG
Name prefix lists	DRUG:	All the variables that begin with "DRUG"
Special SAS name lists	_NUMERIC_	All numeric variables already defined in the current DATA step
	CHARACTER	All character variables already defined in the current DATA step
	ALL	All variables already defined in the current DATA step

1.5.1 Selecting Variables with the KEEP=
Data Set Option or KEEP Statement

To create a data set by selecting a number of variables, you can use either the KEEP= data set option or the KEEP statement. Using data set options allows you to specify certain actions that apply to either the input or output data sets. To use the data set options, you need to place the data set options in parentheses after the data set name. If you need to specify more than one option, you need to separate them with spaces. The general syntax for data set options is as follows:

```
(option-1=value-1<...option-n=value-n>)
```

You can select the variables that you would like to keep by using the KEEP= data set option to control either the input data set or the output data set. The KEEP = data set option has the following form:

```
KEEP=variable-list
KEEP=variable-1 <...variable-n>
```

In the KEEP= option, you can either list the variables that you would like to keep individually (separated by a space) or use the *variable-list* notation.

Specifying the KEEP= data set option after the data set name in the SET statement reads only the specified variables from the input data set. Specifying the KEEP= data set option in the DATA statement controls which variables are written to the output data set. For example, Program 1.11 creates a data set by reading the variables ID, SMOKE, and AGE from the HEARING data set. Using the KEEP= option after DAT1 in the DATA statement will yield the same result; however, specifying the KEEP= option in the SET statement in this example is more efficient because SAS reads only the desired variables.

Program 1.11:

```
data dat1;
    set hearing(keep = id smoke age);
run;
```

The easiest way to verify whether the DAT1 data set was created correctly is to check your SAS log. Based on the SAS log from Program 1.11, you can see that DAT1 contains 34 observations and 3 variables. You can also use PROC CONTENTS to examine the contents of the newly created data set.

Log from Program 1.11:

```
136 data dat1(keep = id smoke age);
137     set hearing;
138 run;

NOTE: There were 34 observations read from the data set
      WORK.HEARING.
NOTE: The data set WORK.DAT1 has 34 observations and 3
variables.
NOTE: DATA statement used (Total process time):
      real time               0.07 seconds
      cpu time                0.01 seconds
```

Alternatively, you can select variables in the DATA step by using the KEEP statement, which has the following form:

```
KEEP variable-list;
KEEP variable-1 <...variable-n>;
```

Notice that there will be no equal sign after the KEEP keyword in the KEEP statement. Program 1.12 creates the same data set by using the KEEP statement instead of the KEEP= data set option.

Program 1.12:

```
data dat1;
    set hearing;
    keep id smoke age;
run;
```

1.5.2 Selecting Variables with the DROP= Data Set Option or DROP Statement

You can also use the DROP= data set option or DROP statement to select variables that you want to remove when creating a data set. The DROP= data set option has the following form:

```
DROP=variable-list
DROP=variable-1 <...variable-n>
```

Program 1.13 selects the same variables (ID, SMOKE, and AGE) from the HEARING data set by using the DROP= option.

Program 1.13:

```
data dat2;
    set hearing(drop = race preg -- income);
run;
```

In Program 1.13, PREG -- INCOME (a named range list) is equivalent to listing all the variables between PREG and INCOME.

You can also use the DROP statement to select variables, which has the following form:

```
DROP variable-list;
DROP variable-1 <...variable-n>;
```

Program 1.14 achieves the same result by using the DROP statement.

Program 1.14:

```
data dat2;
    set hearing;
    drop race preg -- income;
run;
```

If you have more variables to drop than you have to keep, it will be easier for you to use the KEEP= option or the KEEP statement to save you some typing. Conversely, if you have more variables to keep than to drop, you should use the DROP= option or the DROP statement.

1.5.3 Where to Specify the DROP= and KEEP= Data Set Options and DROP/KEEP Statements

You can specify the DROP= and KEEP= data set options in either the DATA statement or the SET statement after the name of the SAS data set, depending upon whether or not you want to process values of the variables in the DATA step. For example, Program 1.15 creates a data set that contains only two variables, ID and INCOME_HI, which is used to indicate whether income level is above or below $50,000.

Program 1.15:

```
data dat3 (drop = income);
    set hearing (keep = id income);
    if income > 50000 then income_hi = 1;
    else income_hi = 0;
run;
```

In Program 1.15, the KEEP= option is used in the SET statement to read the ID and INCOME variables from the input data HEARING. The INCOME variable is required because you need to use the INCOME variable in the DATA step to create a new variable (INCOME_HI) by using the IF-THEN/ELSE statement. Because you don't need the INCOME variable in the output data set, you can then specify the INCOME variable in the DROP= option in the DATA statement.

In DATA steps, the DROP= and KEEP= data set options can be used in either the SET statement to apply to the input data set or the DATA statement to apply to the output data set. However, the KEEP and DROP statements apply only to output data sets. In DATA steps, when you create multiple output data sets, you can use the DROP= or KEEP= data set options to write different variables to different data sets. The DROP or KEEP statements apply to all output data sets. For example, Program 1.16 uses one DATA step to create two SAS data sets: DAT4 and DAT5. DAT4 is created by keeping only the ID, RACE, and SMOKE variables, and DAT5 is created by keeping only the ID, AGE, and PREG variables. The KEEP= option is used to control which variables are retained after each data set name.

Program 1.16:

```
data dat4 (keep = id race smoke)
     dat5 (keep = id age preg);
    set hearing;
run;
```

You can use the DROP= or KEEP= data set options only in PROC steps, not in the DROP and KEEP statements.

1.6 Changing the Appearance of Data

When you encounter a new data set, you will often notice that the variable names are frequently very brief. Shorter but more meaningful variable names are often a preferred format because they are easy to type and save storage space. This idea also applies to variable values. For example, the numerical value 1 is often used to represent "yes" and 0 is used to represent "no" for some categorical variables.

Labeling variable names and formatting variable values are often conducted by SAS programmers. The purpose of doing so is to provide a more appealing look when printing the output from a procedure. Labeling and formatting variables only modify the appearance of the variables' names and values; these actions have no effect whatsoever on the variables' original names and values.

1.6.1 Labeling Variables

To create a descriptive label for a variable, you can use the LABEL statement, which has the following form:

```
LABEL variable-1=label-1... <variable-n=label-n>;
```

You can label one or more variables within one LABEL statement and separate them by space(s). The *variable(s)* in the LABEL statement are the names of the variables that you want to label and *label(s)* are their corresponding labels. The assigned labels can contain blanks and can be no more than 256 characters. If the assigned label contains semicolons (;) or equal signs (=), you can enclose the label in either single or double quotations. If the label contains a single quote ('), you must enclose the labels in double quotations.

To remove labels from variables, you need to use the LABEL statement and associate the variables with a single blank space in quotation marks. Here's the syntax for removing labels:

```
LABEL variable-1=' '... <variable-n=' '>;
```

You can use the LABEL statement in either the DATA step or PROC step. When using a LABEL statement in a DATA step, you associate labels with the variables permanently; in this situation, the assigned label becomes one of the variable's attributes. When using a LABEL statement in the PROC step, the assigned label is available only to the output that the current procedure generated. The LABEL statement in the PROC step does not add permanent labels in the input data set except when the LABEL statement is used with the MODIFY statement in the DATASETS procedure.

Program 1.17 creates a new data set, HEARING1_1. In the HEARING1_1 data set, the variables HEARING and INCOME are assigned with permanent labels. When using PROC PRINT to display the data set contents, in order to see variable labels in the output, you need to use the LABEL option in the PROC PRINT statement. The OBS= option is used to specify the last observation that SAS needs to process the input data set. Thus, using OBS=5 restricts the printed output to the first five observations. Similarly, you can use the FIRSTOBS= option to specify the first observation that SAS processes in an input data set.

Program 1.17:

```
data hearing1_1;
    set hearing;
    label hearing = Hearing Loss
          income = "People's Income";
run;

title 'Assigning Labels Permanently';
proc print data = hearing1_1(obs = 5) label ;
    var hearing income;
run;

proc contents data = hearing1_1;
run;
```

Output from Program 1.17:

```
              Assigning Labels Permanently
                     Hearing       People's
           Obs        Loss          Income
            1          no            29000
            2                        28700
            3                        59000
            4          no           120000
            5                        29000
```

Partial Output from PROC CONTENTS from Program 1.17:

```
          Alphabetic List of Variables and Attributes
          #    Variable    Type    Len    Label
          4    Age         Num      8
          6    Hearing     Char     3     Hearing Loss
          7    Income      Num      8     People's Income
          5    Preg        Num      8
          1    id          Char     4
          2    race        Char     1
          3    smoke       Char     7
```

Based on the partial output from PROC CONTENTS from Program 1.17, you can see that the HEARING and INCOME variables have permanent labels that are under the LABEL attribute column.

Program 1.18 calculates mean, median, and standard deviation for the INCOME variable by each category of ethnicity. In PROC MEANS, the LABEL statement is used to assign the RACE variable a temporary label,

"Ethnicity." Notice that the label attribute for the INCOME variable is listed in the output because the INCOME variable is assigned a permanent label in the DATA step in Program 1.17.

Program 1.18:

```
title 'Assign temporary label to RACE variable';
proc means data = hearing1_1 mean median std;
    label race = Ethnicity;
    class race;
    var income;
run;
```

Output from Program 1.18:

```
              Assign temporary label to RACE variable
                      The MEANS Procedure
            Analysis Variable : Income People's Income
Ethnicity      N Obs        Mean        Median       Std Dev
- - - - - - - - - - - - - - - - - - - - - - - - - - - - - - -
A                5       61420.00      39100.00      45499.25
B                5       61200.00      39100.00      44726.45
H                4       44800.00      37000.00      21332.92
W               20       53575.00      48550.00      25353.54
- - - - - - - - - - - - - - - - - - - - - - - - - - - - - - -
```

1.6.2 Formatting Variable Values Using SAS FORMATS

A *format* is an instruction to tell SAS how to write data values. You can use formats to control the written appearance of data values in the output. Variable values can be formatted by using either SAS formats or user-defined formats that are created from the FORMAT procedure. This section covers only the SAS format; the PROC FORMAT is covered in Chapter 10. To associate formats with variables, you need to use the FORMAT statement, which has the following form:

```
FORMAT variable-1 <... variable-n> format
       variable-1 <... variable-n> format;
```

In the FORMAT statement, you associate a *format* with one or more *variables* that are listed before the specified *format*. That is to say, you can associate the same format with multiple variables in a single FORMAT statement. You can also associate different formats with different variables.

SAS has a large selection of formats that allow you to format character and numeric variables. To format standard character data, you will use the following format:

```
$w.
```

The dollar sign ($) is required for formatting character values. The *w* field is used to specify total width of the character values. The period (.) is a required component of the format.

There are many formats to write nonstandard character values. The difference between standard and nonstandard character formats is that nonstandard character formats contain a keyword in between the dollar sign and the *w* field. For example, the *$UPCASEw.* format converts character values to uppercase when writing the data value, the *$QUOTEw.* format writes data values enclosed in double quotation marks, etc.

To write numeric values in the standard format, you need to use the following form:

w.d

The first field *w* is used to specify total width of the numerical values, including the decimal point. The second field is a period (.). The last field *d* is optional; it is used to specify the number of digits to the right of the decimal point in the numeric value. If *d* is omitted, *w.d* format writes the value without a decimal point.

A nonstandard numeric format contains a keyword in front of the *w* field. For example, the *DOLLARw.d* format will write a number as follows: a leading dollar sign in the starting position, a comma that separates every three digits, and a period that separates the decimal fraction. When using the *DOLLARw.d* format, the *w* field must be large enough to include the dollar sign, comma, and the decimal points.

If you associate a variable(s) with a format in the DATA step, the associated format will become permanent and the format will become the variable's attribute. Like the LABEL statement, if you use a FORMAT statement in some PROC steps, the associated format will only be available for the output of the current procedure.

To disassociate a format from a variable, use the variable in a FORMAT statement without specifying any formats in a DATA step. Here is the syntax to disassociate a format from a variable:

FORMAT variable-1 <... variable-n>;

Program 1.19 associates the SMOKE and HEARING variables with the $UPCASE5. format and the INCOME variable with the DOLLAR11.2 format. Because the width of the $UPCASE5. format is not long enough, only the first five characters in the SMOKE variable are printed in the output.

Program 1.19:

```
data hearing1_2;
    set hearing;
    format smoke hearing $upcase5. income dollar11.2;
run;
```

```
title 'Assigning formats to variable';
proc print data = hearing1_2(firstobs = 6 obs = 12);
    var smoke hearing income;
run;
```

Output from Program 1.19:

```
          Assigning formats to variable
     Obs     smoke     Hearing         Income
      6      CURRE                 $19,000.00
      7      CURRE                 $23,900.00
      8      CURRE                 $39,000.00
      9      CURRE                 $39,100.00
     10      NEVER       NO        $48,000.00
     11      NEVER       NO        $30,000.00
     12      NEVER                 $25,000.00
```

Exercises

Exercise 1.1. Create a SAS data set that is based on HEARING.SAS7BDAT. The information about downloading the testing data sets can be found in the Preface.

1. In this data set, create a variable M_INCOME (monthly income) based on the variable INCOME (yearly income).
2. Create another variable, HEARING_INFO, which is based on the variable HEARING. If the HEARING variable contains missing values, HEARING_INFO will be assigned with value 1; otherwise, HEARING_INFO will be assigned with value 0.
3. Label the variable M_INCOME with "monthly income".
4. Format the M_INCOME variable with the FRACT9. format. Note: You might need to check the SAS documentation for this format. If you have difficulty finding this document, you can search "Fractw. Format SAS" using your preferred online search engine.
5. In the resulting data set, you need to keep the ID, HEARING, HEARING_INFO, INCOME, and M_INCOME variables.

Once this data set is created, use PROC FREQ to create a two-way contingency table for the variables HEARING and HEARING_INFO to confirm that HEARING_INFO was created correctly. Explore the NOPERCENT, NOCOL, and NOROW options for this procedure.

TABLE 1.5

Variable Information for EX1_3.DAT

Variable Name	Description	Locations	Variable Type
ID	Subject's ID	Columns 1–4	Character
GENDER	Gender	Columns 6–11	Character
SMOKE	Smoking status	Columns 13–15	Character
AGE	Age	Columns 17–18	Numeric
DISEASE	Disease status	Column 20	Numeric

Exercise 1.2. The MEANS procedure is introduced in Section 1.4.5. There are other statements in this procedure that were not presented in this section. For example, you can use the OUTPUT statement to output the calculated statistics to a SAS data set. The detailed explanation of this statement can be found in the SAS documentation ("The MEANS Procedure" article). For this exercise, create a SAS data set, HEARING_STATS, via the OUTPUT statement. The HEARING_STATS data set will contain the minimum, first quartile (Q1), median, third quartile (Q3), and maximum values for the AGE and INCOME variables by each category of the RACE variable in the HEARING data set. Because the requested statistics are outputted to a SAS data set, you can use the NOPRINT option in the PROC MEANS statement to suppress the output being listed in the output window.

Exercise 1.3. Create two SAS data sets by using one DATA step. These two SAS data sets are created by reading the raw data set EX1_3.DAT. The description of the data set is listed in Table 1.5. One of the SAS data sets that you are creating will contain only the ID, GENDER, and SMOKE variables; the other one will contain only the ID, AGE, and DISEASE variables.

2

Creating Variables Conditionally

2.1 The IF-THEN/ELSE Statement

Creating variables in the DATA step is an essential task in DATA step programming. If the assigned values to the newly created variable are the same across all the observations, the variable can be created by using the assignment statement. However, in most applications, assigned values are often different across the records, and these values are generated depending upon a certain condition. In this situation, a common approach to creating a variable conditionally is to use the IF-THEN/ELSE statement.

2.1.1 Steps for Creating a Variable

When creating a new variable, novice programmers tend to simply create the variables without checking the accuracy of the variables. A proper way to create a variable based on an existing variable should consist of three important steps:

1. Evaluating the existing variable
2. Creating the new variable
3. Checking the accuracy of the newly created variable

Using these three steps is especially important when creating variables conditionally, because assigned values for the newly created variables are not the same across the observations. You have to make sure that each observation acquires its intended value.

Numerous SAS® procedures can be used to evaluate either the existing or the newly created variables. For example, PROC PRINT will give you a general idea what the data looks like. Instead of printing the entire data set, you can use the OBS= data set option to print only the first few observations of the data set. To examine the variable attributes and the total number of observations, you can use PROC CONTENTS. You can use PROC MEANS to evaluate the distribution of numeric variables; using the NMISS option in PROC MEANS is especially useful for examining missing numerical values.

To examine the frequency of a categorical variable, one often uses PROC FREQ. You can use the MISSING or MISSPRINT options in PROC FREQ to check whether a categorical variable contains missing values.

Suppose that you would like to create an indicator variable (AGE_HI) based on the numeric AGE variable in the HEARING data set created in Chapter 1. You would like to assign a value of 1 to AGE_HI when AGE is greater than its median value; otherwise AGE_HI is set to zero (0).

To follow the three-step procedure above, you need to examine the AGE variable first. In addition to knowing the median value of the AGE variable, it is important to know the number of missing and non-missing values for the AGE variable because if the AGE variable is missing for an observation, AGE_HI should be assigned with a missing value as well. All this information can be found from PROC MEANS by using the MEDIAN (median value), N (number of non-missing values), and NMISS (number of missing values) options.

Program 2.1:

```
title 'Evaluate the AGE variable';
proc means data = hearing n nmiss median maxdec = 2;
     var age;
run;
```

Output from Program 2.1:

```
      Evaluate the AGE variable
      The MEANS Procedure
      Analysis Variable : Age
                    N
         N    Miss            Median
       _ _ _ _ _ _ _ _ _ _ _ _ _ _
        33       1             26.00
       _ _ _ _ _ _ _ _ _ _ _ _ _ _
```

The output from PROC MEANS from Program 2.1 shows that the median age is 26. There are 33 non-missing values and one missing value for AGE. In the second step (Program 2.2), you can create the variable AGE_HI by using the IF-THEN/ELSE statement, which was introduced in Chapter 1. The syntax is also listed below:

```
IF expression THEN statement;
<ELSE statement;>
```

Program 2.2:

```
data hearing2_1;
     set hearing;
```

```
    if age > 26 then age_hi = 1;
    else age_hi = 0;
run;
```

Once the AGE_HI variable is created, the final (and most important) step is to validate that the newly-created variable (AGE_HI) is indeed correctly created. To check the accuracy of AGE_HI, you need to make sure that the maximum value for AGE within the low AGE_HI group (AGE_HI = 0) is 26 and the minimum value of AGE within the high AGE_HI group (AGE_HI = 1) is greater than 26. Program 2.3 uses PROC MEANS to examine the minimum and the maximum values of AGE by each category of AGE_HI. Furthermore, Program 2.3 also examines the number of missing and non-missing values within each level of AGE_HI by using the N and NMISS options.

Program 2.3:

```
title 'Checking AGE_HI is created correctly';
proc means data = hearing2_1 n nmiss min max maxdec = 2;
    class age_hi;
    var age;
run;
```

Output from Program 2.3:

```
              Checking AGE_HI is created correctly
                      The MEANS Procedure
                    Analysis Variable : Age
                    N                 N
   age_hi        Obs      N        Miss     Minimum      Maximum
 - - - - - - - - - - - - - - - - - - - - - - - - - - - - - - - - -
        0         18      17         1        15.00        26.00
        1         16      16         0        28.00        36.00
 - - - - - - - - - - - - - - - - - - - - - - - - - - - - - - - - -
```

Based on the output from Program 2.3, the AGE range is correct when AGE_HI equals 1. There is a problem, however, when AGE_HI is set to zero in the IF-THEN/ELSE statement. Since missing (.) is a number, in fact the smallest numeric value in SAS, the *missing* value for AGE incorrectly translates to a *nonmissing* zero in AGE_HI.

2.1.2 Handling Missing Values When Creating Variables

In most situations, missing values in the source variable should translate into missing values for the target variable. In the previous example involving AGE_HI, the IF-THEN/ELSE statement should have excluded missing values for AGE.

One way to examine the observations that may contain numerical missing values, which are denoted with periods (.), is to check variables with the EQ comparison operator. For example,

```
if AGE EQ . then... ;
```

If you want to compare a character variable with a missing value, you need to compare the character variable with a blank space enclosed in either single or double quotation marks. Alternatively, you can use the MISSING function to check whether its argument contains any missing values. The MISSING function has the following form:

MISSING(*numeric-expression* | *character-expression*)

The advantage of using the MISSING function is that you don't need to know whether its argument is character or numeric. The argument in the MISSING function can be constant, variable, or an expression. If the argument contains a missing value, the MISSING function will return a value of 1; otherwise, it will return 0. Thus, you can create the AGE_HI variable by excluding observations with missing AGE observations by writing the following statement:

```
if missing(age) eq 0 then
    if age > 26 then age_hi = 1; else age_hi = 0;
```

In the SAS statement above, the IF-THEN/ELSE statement that created the AGE_HI variable is nested within the THEN clause of the *outer* IF-THEN statement. The observations for the AGE_HI variable are assigned with either a 1 or 0 only for observations where AGE is not missing. Program 2.4 correctly creates the AGE_HI variable.

Program 2.4:

```
data hearing2_2;
    set hearing;
    if missing(age) eq 0 then
        if age > 26 then age_hi = 1; else age_hi = 0;
run;

title 'Creating AGE_HI considering the missing value';
proc means data = hearing2_2 n nmiss min max maxdec = 2;
    class age_hi;
    var age;
run;
```

Output from Program 2.4:

```
        Creating AGE_HI considering the missing value
        The MEANS Procedure
```

```
              Analysis Variable : Age
                N                  N
    age_hi     Obs      N        Miss    Minimum    Maximum
-  -  -  -  -  -  -  -  -  -  -  -  -  -  -  -  -  -  -  -  -  -

       0       17       17         0       15.00      26.00
       1       16       16         0       28.00      36.00
-  -  -  -  -  -  -  -  -  -  -  -  -  -  -  -  -  -  -  -  -  -
```

Notice that PROC MEANS in Program 2.4 correctly shows that there are no
missing values in both the 0 and 1 strata for the AGE_HI variable. However,
the person with the missing AGE was not shown in the output. In order to
show the missing values that have been excluded from the output, you need
to use the MISSING option in the PROC MEANS statement (see Program 2.5).
The MISSING option will consider missing values as valid values to create
the combinations of class variables.

Program 2.5:

```
title 'Use the MISSING option to show missing values';
proc means data = hearing2_2 n nmiss min max maxdec = 2 missing;
    class age_hi;
    var age;
run;
```

Output from Program 2.5:

```
       Use the MISSING option to show missing values
                 The MEANS Procedure
              Analysis Variable : Age
                N                  N
    age_hi     Obs      N        Miss    Minimum    Maximum
-  -  -  -  -  -  -  -  -  -  -  -  -  -  -  -  -  -  -  -  -  -

       .        1        0         1        .          .
       0       17       17         0       15.00      26.00
       1       16       16         0       28.00      36.00
-  -  -  -  -  -  -  -  -  -  -  -  -  -  -  -  -  -  -  -  -  -
```

2.1.3 TRUE and FALSE: Logical Expressions

The *expression* in the IF-THEN/ELSE statement often contains a comparison
operator, such as EQ, to perform a comparison. The execution of the THEN
or the ELSE clause depends upon the value that results from the comparison. In the last example, the value from the MISSING function is compared
with 0 by using the EQ operator. If the comparison is evaluated to be TRUE,
then the value for the AGE_HI variable is generated. Alternatively, you can

compare the value from the MISSING function with 1 by adding the NOT operator. For example,

```
if not (missing(age) eq 1) then
    if age > 26 then age_hi = 1; else age_hi = 0;
```

In SAS, any numerical value other than 0 or the missing value is considered TRUE, and the missing value and 0 are considered FALSE. Since the MISSING function returns a numerical value with either 1 (for TRUE) or 0 (for FALSE), it is redundant to compare values from the MISSING function with 1 explicitly. You can simply write the following:

```
if not missing(age) then
    if age > 26 then age_hi = 1; else age_hi = 0;
```

In some applications, you also need to compare a numeric variable explicitly with 1. For example, in the HEARING data set, the variable PREG is coded with 1 for indicating being pregnant and 0 for not being pregnant. Suppose that you would like to assign pregnant women to group "A" and nonpregnant women to group "B".

Program 2.6 creates the GROUP variable that is based on the PREG variable. All three steps of creating variables are included in the program. During the first step, the variable PREG is evaluated with PROC FREQ instead of PROC MEANS. PROC FREQ is preferred because PREG has just two distinct values (0 and 1), and frequencies need to be examined at each level.

In the DATA step, the statement "IF PREG THEN..." is equivalent to "IF PREG = 1 THEN...". Once the GROUP variable is created from the DATA step, PROC FREQ returns as a third step to check if GROUP is correctly created by examining the frequency in the cross-tabulation of the PREG and GROUP variables.

Program 2.6:

```
/* Step # 1 */
title 'Check PREG variable';
proc freq data = hearing;
    tables preg/missing nocol nopercent;
run;

/* Step # 2 */
data hearing2_3;
    set hearing;
    if not missing(preg) then
        if preg then group = "A";
        else group = "B";
run;

/* Step # 3 */
title 'Check TRIAL is created correctly';
```

```
proc freq data = hearing2_3;
    tables preg*group/missing norow nocol nopercent;
run;
```

Output from Program 2.6:

```
                    Check PREG variable
                    The FREQ Procedure
                    Preg      Frequency
                    — — — — — — — — —

                       .            4
                       0           19
                       1           11

              Check TRIAL is created correctly
                    The FREQ Procedure
                  Table of Preg by group
          Preg        group
        Frequency,        |A      |B       | Total
        — — — — -+— — — +— — — +— — — +
               . |    4 |    0 |    0 |     4
        — — — — -+— — — +— — — +— — — +
             0 |    0 |    0 |   19 |    19
        — — — — -+— — — +— — — +— — — +
             1 |    0 |   11 |    0 |    11
        — — — — -+— — — +— — — +— — — +
          Total      4      11     19     34
```

A classic logic error often occurs when using the OR operator in the IF-THEN/ELSE statement. For example, "if foo = 10 or 20" always resolves to TRUE because 20 is treated as a single expression and evaluated as TRUE (you actually intended to write "if foo = 10 or foo = 20"). This type of error is often difficult to detect because the syntax is completely correct.

2.1.4 The LENGTH Attribute

In the previous example, values for GROUP have the same one-character length ("A" or "B"). When creating a character variable that contains values with different lengths, you should preset the variable length before creating the variables; otherwise, the values with longer lengths might be truncated. For example, suppose that you want to create a character variable (AGE_CAT) based on the AGE variable in the HEARING data set.

You would like to assign AGE_CAT to "old" when AGE is greater than its median and "young" when AGE is less than the median. At first glance, you might think that the code in Program 2.7 would work.

Program 2.7:

```
data hearing2_4;
    set hearing;
```

```
    if not missing(age) then
        if age > 26 then age_cat = "old";
        else age_cat = "young";
run;

title 'The first 5 observations of HEARING2_4 data set';
proc print data = hearing2_4(obs = 5);
    var age age_cat;
run;
```

Output from Program 2.7:

```
        The first 5 observations of HEARING2_4 data set
             Obs    Age     age_cat
              1      26        you
              2      26        you
              3      32        old
              4      32        old
              5      34        old
```

Notice that the value "young" is truncated to "you." The reason for the truncation is that when creating character variables, SAS will allocate the same number of bytes of storage space as there are characters in the first value that it encounters in the data set for that variable. In the program above, SAS allocates 3 bytes because "low" was encountered first. To avoid mistakes caused by truncation, *always* use the LENGTH statement described below to define character variables in a data step:

```
LENGTH variable(s) <$> length;
```

You can specify one or more *variables* in a LENGTH statement. The dollar sign ($) in the LENGTH statement is used to apply character variables. In the LENGTH statement, *length* is the number of bytes that are used for storing variable values. The length for numeric variables is up to 8 bytes, and the length for character variables is up to 32,767 bytes. The LENGTH statement must be placed before any other reference to the variable in the DATA step; otherwise it won't take effect. Program 2.8 correctly creates the AGE_CAT variable by placing the LENGTH statement before the IF-THEN statement.

Program 2.8:

```
data hearing2_5;
    length age_cat $ 5;
    set hearing;
    if not missing(age) then
        if age > 26 then age_cat = "old";
        else age_cat = "young";
run;
```

```
title 'The first 5 observations of HEARING2_5 data set';
proc print data = hearing2_5(obs = 5);
    var age age_cat;
run;
```

Output from Program 2.8:

```
       The first 5 observations of HEARING2_5 data set
           Obs     Age     age_cat
            1       26      young
            2       26      young
            3       32      old
            4       32      old
            5       34      old
```

2.1.5 DO Group

Sometimes you might find it necessary to execute a group of statements as one unit, which can be accomplished by using the DO statement. The DO statement has the following form:

```
DO;
      SAS statement1
...
      SAS statementn
END;
```

The SAS statements between the DO and END statements are called a DO group. You can nest DO groups within DO groups. A DO group is often used within IF-THEN/ELSE statements. Suppose that you would like to create two variables: PREG_INFO and PREG_SMOKER. These two variables are created by the following conditions:

- PREG_INFO is assigned to
  - 1 if the PREG variable is not missing
  - 0 if PREG is missing
- PREG_SMOKER is assigned to
  - 1 if PREG is not missing and the SMOKE variable equals "current"
  - 0 if PREG is not missing and SMOKE is not "current"

Since both the PREG_INFO and PREG_SMOKER variables require a check to see if PREG is missing, you can utilize a DO group with an IF-THEN/ELSE statement to create these two variables.

Program 2.9 starts with the FREQ procedure before creating the variables PREG_INFO and PREG_SMOKER. Based on the output, you can see that

there is only one pregnant smoker and four people have missing values for
the PREG variable. When creating PREG_INFO and PREG_SMOKER in the
DATA step, a DO group is used to assign PREG to 1 and to create the PREG_
SMOKER variable for the condition when PREG is not missing. The final step
uses PROC FREQ again to verify whether these two variables are created
correctly.

Program 2.9:

```
title 'Frequency Tables: Preg by Smoke';
proc freq data = hearing;
    tables preg*smoke/missing norow nocol nopercent;
run;


data hearing2_6;
    set hearing;
    if not missing(preg) then
    do;
        preg_info = 1;
        if smoke = "current" and preg = 1 then preg_smoker = 1;
        else preg_smoker = 0;
    end;
    else preg_info = 0;
run;


title 'Check if PREG_SMOKER and PREG_INFO are created cor-
rectly';
proc freq data = hearing2_6;
    tables preg_smoker preg_info/missprint;
run;
```

Output from Program 2.9:

Frequency Tables: Preg by Smoke

```
                        The FREQ Procedure
                      Table of Preg by smoke
        Preg       smoke
        Frequency|      | current |  never  |  past  | Total
        - - - - -+ - - +- - - - - +- - - - -+ - - - - +
               . |  0  |       1  |      1  |     2  |   4
        - - - - -+ - - +- - - - - +- - - - -+ - - - - +
             0 |  0  |       6  |      9  |     4  |  19
        - - - - -+ - - +- - - - - +- - - - -+ - - - - +
             1 |  1  |       1  |      8  |     1  |  11
        - - - - -+ - - +- - - - - +- - - - -+ - - - - +
           Total     1          8         18        7     34
```

```
      Check if PREG_SMOKER and PREG_INFO are created correctly
                      The FREQ Procedure
                                      Cumulative  Cumulative
     preg_smoker    Frequency    Percent    Frequency    Percent
   - - - - - - - - - - - - - - - - - - - - - - - - - - - - - -
            .            4          .            .           .
            0           29        96.67          29         96.67
            1            1         3.33          30        100.00
                  Frequency Missing = 4

                                      Cumulative  Cumulative
     preg_info     Frequency    Percent    Frequency    Percent
   - - - - - - - - - - - - - - - - - - - - - - - - - - - - - -
            0            4        11.76           4         11.76
            1           30        88.24          34        100.00
```

2.2 Executing One of Several Statements

The IF-THEN/ELSE statement in the previous section contains only one optional ELSE clause. With only one ELSE statement, you can execute statements based on two conditions. In the situation where you would like to execute multiple statements that are based on more than two conditions, you can either add multiple ELSE IF clauses to the IF-THEN/ELSE statement or use the SELECT statement.

2.2.1 Multiple IF-THEN/ELSE Statements

Multiple IF-THEN/ELSE statements have the following form:

```
IF expression THEN statement;
ELSE IF expression THEN statement;
<...
ELSE IF expression THEN statement;
<ELSE statement;>>
```

The last optional ELSE clause becomes a default that is automatically executed for all observations failing to satisfy any of the previous IF statements.

Suppose you would like to create a variable (AGEGROUP) based on the AGE variable in the HEARING data set. The AGEGROUP variable will be assigned with values 1, 2, or 3 based on the following conditions:

- AGEGROUP is assigned to 1 for AGE ≤ 20
- AGEGROUP is assigned to 2 for $20 < \text{AGE} \leq 30$
- AGEGROUP is assigned to 3 for AGE > 30

Program 2.10 creates AGEGROUP in two different ways. In the first method, AGEGROUP1 is created by comparing AGE with 30, 20, and the missing value (in descending order) with the > comparison operator. When AGE has a value greater than 30, AGEGROUP1 is set to 3, and the remaining conditions are not evaluated. When AGE is greater than 20 but less than or equal to 30, the AGE > 30 condition is checked again but this time it fails, so processing moves on to the next check. This time there is a hit, because AGE > 20 is satisfied. Now AGEGROUP1 is set to 2. Finally, when AGE is less than or equal to 20, all three conditions have to be checked before a value of 1 can be assigned to AGEGROUP1.

When writing multiple IF-THEN/ELSE statements, it is efficient to compare the threshold value in either ascending or descending order. When comparing the threshold values in descending order, either the > or >= operators should be used. On the other hand, when comparing the threshold values in ascending order, you should use either the < or < = operators. In the second method, AGEGROUP2 is created by comparing AGE with 20 and 30 (in ascending order) with the <= comparison operator after excluding the missing values.

Program 2.10:

```
data hearing2_7;
    set hearing;

    *method1;
    if age > 30 then agegroup1 = 3;
    else if age > 20 then agegroup1 = 2;
    else if age >. then agegroup1 = 1;

    *method2;
    if not missing(age) then
        if age <= 20 then agegroup2 = 1;
        else if age <= 30 then agegroup2 = 2;
        else agegroup2 = 3;
run;

title 'Check AGEGROUP1 is created correctly';
proc means data = hearing2_7 missing n nmiss min max maxdec = 2;
    class agegroup1;
    var age;
run;

title 'Check AGEGROUP2 is created correctly';
proc means data = hearing2_7 missing n nmiss min max maxdec = 2;
    class agegroup2;
    var age;
run;
```

Output from Program 2.10:

```
                 Check AGEGROUP1 is created correctly
                         The MEANS Procedure
                      Analysis Variable : Age
                    N                    N
 agegroup1    Obs        N      Miss      Minimum        Maximum
 - - - - - - - - - - - - - - - - - - - - - - - - - - - - - - - - -
         .     1        0        1          .              .
         1    10       10        0        15.00          20.00
         2    12       12        0        23.00          30.00
         3    11       11        0        31.00          36.00
 - - - - - - - - - - - - - - - - - - - - - - - - - - - - - - - - -


                 Check AGEGROUP2 is created correctly
                         The MEANS Procedure
                      Analysis Variable : Age
                    N                    N
 agegroup2    Obs        N      Miss      Minimum        Maximum
 - - - - - - - - - - - - - - - - - - - - - - - - - - - - - - - - -
         .     1        0        1          .              .
         1    10       10        0        15.00          20.00
         2    12       12        0        23.00          30.00
         3    11       11        0        31.00          36.00
 - - - - - - - - - - - - - - - - - - - - - - - - - - - - - - - - -
```

Program 2.11 illustrates the use of multiple IF-THEN/ELSE statements with the DO group. This program creates two variables, TRIAL and REQUIREINFO, based on the variable PREG.

Program 2.11:

```
data hearing2_8;
    set hearing;
    length trial $4;
    if preg = 1 then do;
        trial = "A";
        requireInfo = 0;
    end;
    else if preg = 0 then do;
        trial = "B";
        requireInfo = 0;
    end;
    else do;
        trial = "Wait";
        requireInfo = 1;
    end;
run;
```

```
title 'Checking if TRIAL and REQUIREINFO are created correctly';
proc freq data = hearing2_8;
    tables (trial requireInfo)*preg/missing nocol norow
        nopercent;
run;
```

Output from Program 2.11:

```
       Checking if TRIAL and REQUIREINFO are created correctly
                        The FREQ Procedure
                      Table of trial by Preg
          trial          Preg
          Frequency|        . |        0|        1|    Total
          - - - -+- - - -+- - - -+- - - -+
          A        |     0 |      0 |     11 |       11
          - - - -+- - - -+- - - -+- - - -+
          B        |     0 |     19 |      0 |       19
          - - - -+- - - -+- - - -+- - - -+
          Wait     |     4 |      0 |      0 |        4
          - - - -+- - - -+- - - -+- - - -+
          Total          4         19        11           34

                  Table of requireInfo by Preg
          requireInfo        Preg
          Frequency|        . |        0|        1|    Total
          - - - -+- - - -+- - - -+- - - -+
              0 |     0 |     19 |     11 |       30
          - - - -+- - - -+- - - -+- - - -+
              1 |     4 |      0 |      0 |        4
          - - - -+- - - -+- - - -+- - - -+
          Total          4         19        11           34
```

2.2.2 Executing Statements Using the SELECT Group

The SELECT group is an alternative method to executing one of several statements, which has the following form:

```
SELECT <(select-expression)>;
    WHEN-1 (when-expression-1 <..., when-expression-n>)
    statement;
    <... WHEN-n (when-expression-1 <..., when-expression-n>)
    statement;>
    <OTHERWISE statement;>
END;
```

A SELECT group consists of several statements that start with the SELECT statement and end with the END statement. In the SELECT statement, the optional *select-expression* is used to specify any SAS expression that can be evaluated into a single value. Within the SELECT group, it contains at least one WHEN statement, which is used to identify which *statement* is to be executed when a WHEN-condition is met. This WHEN-condition, which is either TRUE or FALSE, depends upon whether the *select-expression* is specified.

When a *select-expression* is specified in the SELECT statement, SAS compares the evaluated results from *select-expression* and *when-expression* and returns a value of TRUE or FALSE. If the comparison is true for a WHEN statement, the corresponding *statement* is executed; otherwise, a comparison is performed for either the next *when-expression* within the current WHEN statement or the one in the next WHEN statement. If there is no WHEN-condition that is TRUE, the OTHERWISE statement is executed if one exists. If all the comparisons are false and there is no OTHERWISE statement, SAS will issue an error message and terminate DATA step execution. If the comparison is TRUE for more than one WHEN statement, only the first WHEN statement is executed.

The execution sequence of the WHEN statement when no *select-expression* is specified is similar to when a *select-expression* is specified. However, when no *select-expression* is specified, only the *when-expression* is evaluated and generates a value of TRUE or FALSE. If it is TRUE for a WHEN statement, the corresponding *statement* is executed.

Executing statements that utilize the SELECT group can also be rewritten by using the multiple IF-THEN/ELSE statements. Using the SELECT group is generally more suitable than IF-THEN/ELSE for statements having multiple conditions that need to be processed.

Program 2.12 uses select groups to create a number of variables. The first SELECT group creates the ETHNIC variable by comparing the values from the RACE variable with two groups of values. If the value from the RACE variable equals either "W" or "H", then ETHNIC will be assigned with the "white" value; if RACE equals "B" or "A", then ETHNIC will be assigned with the "non-white" value. The OTHERWISE statement is not utilized, which is not a recommended practice because if the result of all SELECT-WHEN comparisons is false, SAS will issue an error message. A final FALSE would occur, for example, if RACE is missing or if lowercase characters ("w","h","b","a") are assigned to the variable.

The second SELECT group creates the TRIAL and DRUG variables. The DO group is used to group two assignment statements for each WHEN statement. The OTHERWISE statement is used without using an additional statement after the keyword OTHERWISE. This means that if PREG is other than 1 or 0, the TRIAL and DRUG variables will be assigned missing values.

The third SELECT group creates the GROUP variable. An OTHERWISE statement is used within the SELECT group. Subjects with HEARING other than "yes" and "no" values will be assigned to 3 for the GROUP variable.

The last SELECT group creates the variable HIGHINCOME. In this SELECT group, no *select-expression* is specified. In this situation, the *when-expression* is evaluated for selecting which WHEN statement is to be executed.

Program 2.12:

```
data hearing2_9;
    set hearing;
    length ethnic $ 10;

    select (race);
        when ("W", "H") ethnic = "white";
        when ("B", "A") ethnic = "non-white";
    end;

    select (preg);
        when (1) do;
            trial = "A";
            drug = "Treatment";
        end;
        when (0) do;
            trial = "B";
            drug = "placebo";
        end;
        otherwise;
    end;

    select(hearing);
        when ("yes") group = 1;
        when ("no") group = 2;
        otherwise group = 3;
    end;

    select;
        when (income > 100000) highincome = 1;
        when (income >.) highincome = 0;
        otherwise;
    end;
run;

title 'Check if ETHNIC, TRIAL, DRUG, and GROUP are created
correctly';
proc freq data = hearing2_9;
    tables race*ethnic
            preg*trial
            preg*drug
            hearing*group/norow nocol nopercent missing;
run;
```

```
title 'Check if HIGHINCOME is created correctly';
proc means data = hearing2_9 missing n nmiss min max maxdec = 2;
    class highincome;
    var income;
run;
```

Output from Program 2.12:

```
Check if ETHNIC, TRIAL, DRUG, and GROUP are created correctly
                    The FREQ Procedure
                  Table of race by ethnic
          race        ethnic
          Frequency|non-white|white |Total
          - - - -+- - - - -+ - - +
          A        |        5 |    0 |    5
          - - - -+- - - - -+- - - +
          B        |        5 |    0 |    5
          - - - -+- - - - -+- - - +
          H        |        0 |    4 |    4
          - - - -+- - - - -+- - - +
          W        |        0 |   20 |   20
          - - - -+- - - - -+- - - +
          Total             10     24     34


                 Table of Preg by trial
        Preg         trial
        Frequency|           |A        |B          | Total
        - - - -+- - - - +- - - - +- - - - +
             . |      4 |      0 |        0 |    4
        - - - -+- - - - +- - - - +- - - - +
           0 |      0 |      0 |       19 |   19
        - - - -+- - - - +- - - - +- - - - +
           1 |      0 |     11 |        0 |   11
        - - - -+- - - - +- - - - +- - - - +
        Total         4          11         19        34


                 Table of Preg by drug
        Preg         drug
        Frequency|           |Treatment |placebo | Total
        - - - -+- - - - +- - - - - +- - - - +
             . |      4 |        0 |      0 |    4
        - - - -+- - - - +- - - - - +- - - - +
           0 |      0 |        0 |     19 |   19
        - - - -+- - - - +- - - - - +- - - - +
           1 |      0 |       11 |      0 |   11
        - - - -+- - - - +- - - - - +- - - - +
        Total         4          11         19        34
```

```
Check if ETHNIC, TRIAL, DRUG, and GROUP are created correctly
                    The FREQ Procedure
                 Table of Hearing by group
        Hearing        group
        Frequency|       1|        2|        3| Total
        - - - -+- - - -+- - - -+- - - - +
                |     0 |      0 |     23 |    23
        - - - -+- - - -+- - - -+- - - - +
        no      |     0 |      8 |      0 |     8
        - - - -+- - - -+- - - -+- - - - +
        yes     |     3 |      0 |      0 |     3
        - - - -+- - - -+- - - -+- - - - +
        Total         3        8       23      34


            Check if HIGHINCOME is created correctly
                    The MEANS Procedure
                 Analysis Variable : Income
                   N              N
    highincome    Obs     N      Miss      Minimum         Maximum
    - - - - - - - - - - - - - - - - - - - - - - - - - - - -
             0     31     31        0     13900.00        98000.00
             1      3      3        0    113000.00       134000.00
    - - - - - - - - - - - - - - - - - - - - - - - - - - - -
```

2.3 Modifying the IF-THEN/ELSE Statement with the Assignment Statement

When creating an indicator variable (a variable with values of 1 or 0), you can use the IF-THEN/ELSE statement. For example,

```
if age>26 then age_hi = 1;
else age_hi = 0;
```

In this situation, you can modify the IF-THEN/ELSE statement by using the *assignment* statement, which was introduced in Chapter 1. The assignment statement has the following form:

```
variable=expression;
```

In the assignment statement, *variable* is either a new or existing variable, and *expression* is any valid SAS expression. The IF-THEN/ELSE statement above can be rewritten by utilizing an assignment statement. For example,

```
age_hi = age>26;
```

The expression age>26 is evaluated with 1 for observations with ages greater than 26, and the value 1 will then be assigned to AGE_HI. For observations with ages less than or equal to 26, AGE_HI will be assigned to 0. Similarly, you can also use the assignment statement for creating a variable with ordinal numeric values. For instance, in a previous example, AGEGROUP was created by utilizing the multiple IF-THEN/ELSE statement:

```
if age>30 then agegroup = 3;
else if age>20 then agegroup = 2;
else if age>. then agegroup = 1;
```

The multiple IF-THEN/ELSE statement is equivalent to writing the following assignment statement:

```
agegroup = (age>.) + (age>20) + (age>30);
```

The three parentheses in the assignment statement above are necessary to ensure the three comparisons are evaluated before the addition operation because the addition operator has higher evaluation priority than the comparison operators. Examples for creating the AGEGROUP variable by using different age values are summarized in Table 2.1. The variable AGEGROUP is created in Program 2.13.

Program 2.13:

```
data hearing2_10;
    set hearing;
    agegroup = (age>.) + (age>20) + (age>30);
run;

title 'Check if AGEGROUP is created correctly';
proc means data = hearing2_10 n nmiss min max maxdec = 2
  missing;
    class agegroup;
    var age;
run;
```

Output from Program 2.13:

```
          Check if AGEGROUP is created correctly
                  The MEANS Procedure
                Analysis Variable  : Age
                   N              N
   agegroup       Obs      N     Miss    Minimum      Maximum
 _ _ _ _ _ _ _ _ _ _ _ _ _ _ _ _ _ _ _ _ _ _ _ _ _ _ _ _ _ _

          0        1       0      1         .            .
          1       10      10      0       15.00        20.00
          2       12      12      0       23.00        30.00
          3       11      11      0       31.00        36.00
 _ _ _ _ _ _ _ _ _ _ _ _ _ _ _ _ _ _ _ _ _ _ _ _ _ _ _ _ _ _
```

TABLE 2.1

Examples for Creating the AGEGROUP Variable

When AGE Equals...	(age>.) + (age>20) + (age>30) Evaluates to the Following	agegroup Is Assigned with
. (missing value)	$0 + 0 + 0$	0
15	$1 + 0 + 0$	1
25	$1 + 1 + 0$	2
35	$1 + 1 + 1$	3

Exercises

Exercise 2.1. Using the IF-THEN/ELSE statement, create the following three variables based on the variables in the GRADE.SAS7BDAT data set:

1. MATH_POINT:
   - MATH_POINT is assigned to 4 if variable MATH = "A"
   - MATH_POINT is assigned to 3 if variable MATH = "B"
   - MATH_POINT is assigned to 2 if variable MATH = "C"
   - MATH_POINT is assigned to 1 if variable MATH = "D"
   - MATH_POINT is assigned to 0 if variable MATH = "F"

2. ENGLISH_POINT:
   - ENGLISH_POINT is assigned to 4 if variable ENGLISH ≥ 90
   - ENGLISH_POINT is assigned to 3 if variable 80 ≤ ENGLISH < 90
   - ENGLISH_POINT is assigned to 2 if variable 70 ≤ ENGLISH < 80
   - ENGLISH_POINT is assigned to 1 if variable 60 ≤ ENGLISH < 70
   - ENGLISH_POINT is assigned to 0 if variable ENGLISH < 60 and not missing

3. PE_GRADE:
   - PE_GRADE is assigned to "pass" if variable PE = 1
   - PE_GRADE is assigned to "no pass" if variable PE = 0

Exercise 2.2. Using the SELECT group, create the same three variables in *Exercise 2.1*.

Exercise 2.3. Instead of using the multiple IF-THEN/ELSE statement to create variables MATH_POINT and ENGLISH_POINT in *Exercise 2.1*, use one IF-THEN/ELSE statement with the assignment statement to create this variable following the examples in Section 2.3.

3

Understanding How the DATA Step Works

3.1 DATA Step Processing Overview

A common befuddlement often facing beginning SAS® programmers is that the SAS data set that they create is not what they intended to create—that is, there are more or less observations than intended or the value of the newly created variable is not retained correctly. These types of mistakes occur because new programmers often focus exclusively on language syntax but fail to understand how the DATA step actually works.

The purpose of this chapter is to guide you through how DATA step programming operates, step by step, by way of providing various examples. The material in this chapter is based on a paper that I presented at the Western Users of SAS Software conference (2008).

A DATA step is processed sequentially via the *compilation* and *execution* phases. In the compilation phase, each statement is scanned for syntax errors. If an error is found, SAS will stop processing. The execution phase begins only after the compilation phase ends. Both phases do not occur simultaneously.

In the execution phase, the DATA step works like a loop, repetitively executing statements to read data values and create observations one at a time. Each loop is called an *iteration*. We can refer to this type of loop as the implicit loop, which is different from the explicit loop, by using iterative DO, DO WHILE, or DO UNTIL statements.

Not all SAS statements in the DATA step are executed during the execution phase. Instead, statements in the DATA step can be categorized as *executable* or *declarative*. The declarative statements are used to provide information to SAS and only take effect during the compilation phase. The declarative statements can be placed in any order within the DATA step. Here are a few examples of declarative statements that were covered in the first two chapters:

- LENGTH: setting the internal variable length
- FORMAT: setting the variable output format
- LABEL: defining variable labels

- DROP: indicating which variables are to be omitted in the output file
- KEEP: indicating which variables are to be included in the output file

In contrast to declarative statements, the order in which executable statements appear in the DATA step matters greatly. For example, to read an external text file, you need to start with the INFILE statement, followed by the INPUT statement. The INFILE statement is used to identify the location of the external file, and the INPUT statement instructs SAS how to read each observation. Thus, you must place the INFILE statement before the INPUT statement because SAS needs to know where to find the external file *before* it can read it.

Program 3.1 illustrates how DATA step processing works. This program reads raw data from a text file, EXAMPLE3_1.TXT, and creates one variable, BMI.

EXAMPLE3_1.TXT contains two observations and three variables, NAME (columns 1–7), HEIGHT (columns 9–10), and WEIGHT (columns 12–14). Notice that the WEIGHT variable for the first observation is entered as "12D", which is a data entry error. Since each variable is occupied in a fixed field and the values for these variables are standard character or numerical values, the column input method is best used to read the raw data set.

EXAMPLE3_1.TXT:

```
12345678901234567890123456567890
Barbara  61 12D
John     62 175
```

Program 3.1:

```
data ex3_1;
    infile 'W:\SAS Book\dat\example3_1.txt';
    input name $ 1-7 height 9-10 weight 12-14;
    BMI = 700*weight/(height*height);
    output;
run;
```

Log from Program 3.1:

```
1 data ex3_1;
2      infile 'W:\SAS Book\dat\example3_1.txt';
3      input name $ 1-7 height 9-10 weight 12-14;
4      BMI = 700*weight/(height*height);
5      output;
6 run;
NOTE: The infile 'W:\SAS Book\dat\example3_1.txt' is:
      Filename = W:\SAS Book\dat\example3_1.txt,
      RECFM = V,LRECL = 256,File Size (bytes) = 32,
```

```
        Last Modified = 15Mar2012:09:10:03,
        Create Time = 15Mar2012:09:09:22
NOTE: Invalid data for weight in line 1 12-14.
RULE:─ ─ +─ ─ 1─ ─ +─ ─ 2─ ─ +─ ─ 3─ ─ +─ ─ 4─ ─ +─ ─ 5─ ─
+─ ─ 6
1    Barbara 61 12D 14
name = Barbara height = 61 weight =. BMI =. _ERROR_ = 1 _N_ = 1
NOTE: 2 records were read from the infile 'W:\SAS Book\dat\
example3_1.txt'.
     The minimum record length was 14.
     The maximum record length was 14.
NOTE: Missing values were generated as a result of performing
an operation on missing values.
     Each place is given by: (Number of times) at
(Line):(Column).
     1 at 4:14
NOTE: The data set WORK.EX3_1 has 2 observations and 4
variables.
NOTE: DATA statement used (Total process time):
     real time      0.15 seconds
     cpu time       0.03 seconds
```

3.1.1 DATA Step Compilation Phase

Since Program 3.1 reads in a raw data set, the input buffer is created at the beginning of the compilation phase. The input buffer is used to hold raw data (Figure 3.1). However, if you read in a SAS data set instead of a raw data file, the input buffer will not be created.

SAS also creates the program data vector (PDV) in the compilation phase (Figure 3.1). SAS uses the PDV, a memory area on your computer, to build the new data set. There are two automatic variables, _N_ and _ERROR_, inside the PDV. _N_ equaling 1 indicates the first observation is being processed, _N_ equaling 2 indicates the second observation is being processed, and so on. The automatic variable _ERROR_ is an indicator variable with values of 1 or 0. _ERROR_ equaling 1 signals the data error of the currently processed observation, such as reading the data with an incorrect data type. In addition to the two automatic variables, there is one space allocated for

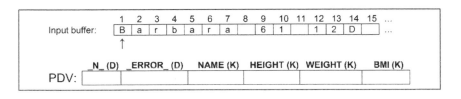

FIGURE 3.1
Input buffer and the program data vector (PDV).

each of the variables that will be created from this DATA step. The variables NAME, HEIGHT, and WEIGHT are read in from an external file, whereas the newly created BMI variable (body mass index) is derived from HEIGHT and WEIGHT.

Notice that some of the variables in the PDV are marked with (D), which stands for "dropped," and others are marked with (K), which stands for "kept." Only the variables marked with (K) will be written to the output data set. Automatic variables, on the other hand, are always marked with a (D) so they are never written out.

During the compilation phase, SAS checks for syntax errors, such as invalid variable names, options, punctuations, misspelled keywords, etc. SAS also identifies the type and length of the newly created variables.

At the end of the compilation phase, the descriptor portion of the SAS data set is created, which includes the data set name, the number of observations, and the number, names, and attributes of variables. The information about the descriptor portion can be generated via the CONTENTS procedure.

3.1.2 DATA Step Execution Phase

At the beginning of the execution phase, the automatic variable _N_ is initialized to 1, and _ERROR_ is initialized to 0 since there is no data error. The nonautomatic variables are set to missing. Once the INFILE statement identifies the location of the input file, the INPUT statement copies the first data line into the input buffer. Then the INPUT statement reads data values from the record in the input buffer according to instructions from the INPUT statement and writes them to the PDV. The values for NAME and HEIGHT are successfully copied from the input buffer to the PDV. However, the value for WEIGHT is "12D", an invalid numeric value that causes _ERROR_ to be set to 1 and WEIGHT to missing. Meanwhile, an error message is sent to the SAS log indicating the location of the data error (see Log from Program 3.1). Next, the assignment statement is executed and BMI will remain missing since operations on a missing value will result in a missing value.

When the OUTPUT statement is executed, only the values from the PDV marked with (K) are copied as a single observation to the output SAS data set, EX3_1. See Figure 3.2 for a detailed explanation of each step in the first iteration.

At the end of the DATA step the SAS system returns to the beginning of the DATA step to begin the next iteration. The values of the variables in the PDV are reset to missing. The automatic variable _N_ is incremented to 2, and _ERROR_ is set to 0. The second data line is read into the input buffer by the INPUT statement. See the illustration in Figure 3.3 for details.

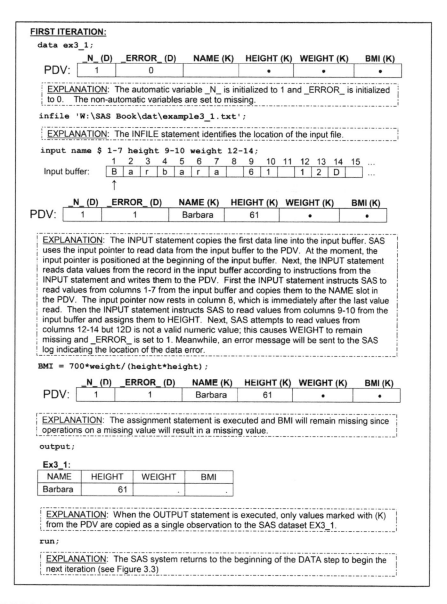

FIGURE 3.2
The first iteration of Program 3.1.

At the end of the DATA step for the second iteration, the SAS system again returns to the beginning of the DATA step to begin the next iteration (see Figure 3.4). The values of the variables in the PDV are reset to missing. The automatic variable _N_ is incremented to 3. SAS attempts to read an observation from the input data set, but it reaches the end-of-file-marker, which means

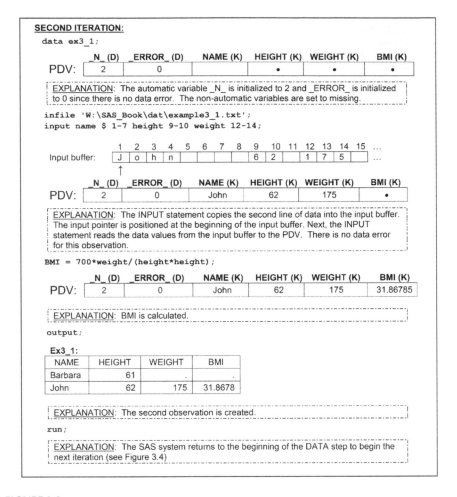

FIGURE 3.3
The second iteration of Program 3.1.

FIGURE 3.4
The third iteration of Program 3.1.

that there are no more observations to read. When an end-of-file marker is encountered, SAS goes to the next DATA or PROC step in the program. (*Program 3.1* is illustrated in *program3.1.pdf*.)

3.1.3 The Importance of the OUTPUT Statement

In Program 3.1, an explicit OUTPUT statement is used to tell SAS to write the current observation from the PDV to a SAS data set immediately. An OUTPUT statement is not actually needed in Program 3.1 because the contents of the PDV are written out by default with an *implicit* OUTPUT statement at the end of the DATA step—exactly where the explicit OUTPUT statement has been placed. In Program 3.2, the explicit OUTPUT statement has been removed so that you can see that the implicit OUTPUT statement works just like its explicit counterpart.

Program 3.2:

```
data ex3_1;
    infile 'W:\SAS Book\dat\example3_1.txt';
    input name $ 1-7 height 9-10 weight 12-14;
    BMI = 700*weight/(height*height);
run;
```

If you place an explicit OUTPUT statement in a DATA step, the explicit OUTPUT statement will override the implicit OUTPUT statement; in other words, once an explicit OUTPUT statement is used to write an observation to an output data set, there is no longer an implicit OUTPUT statement at the end of the DATA step. The SAS system adds an observation to the output data set only when an explicit OUTPUT statement is executed. Furthermore, more than one OUTPUT statement in the DATA step can be used. You will see an example of using multiple OUTPUT statements later in this chapter.

3.1.4 The Difference between Reading a Raw Data Set and a SAS Data Set

When creating a SAS data set based on a raw data set, SAS initializes each variable value in the PDV to missing at the beginning of each iteration of execution, except for the automatic variables, variables that are named in the RETAIN statement, variables that are created by the SUM statement, data elements in a _TEMPORARY_ array, and variables created in the options of the FILE/INFILE statement.

Often an output data set is created based on an existing SAS data set instead of reading in a raw data file. In this instance, SAS sets each variable to missing in the PDV *only* before the first iteration of the execution. Variables will retain their values in the PDV until they are replaced by the new values from the input data set. These variables exist in both the input and output data sets. Often when creating a new data set based on the existing SAS data set, you will create new variables based on existing variables; these new variables are not from the input data set. These new variables will be set to missing in the PDV at the beginning of every iteration of the execution.

3.2 Retaining the Value of Newly Created Variables

In some situations, you may want to retain the values from the newly created variables in the PDV throughout the execution of the DATA step. To prevent the newly created variables from being initialized to *missing* at the beginning of each iteration of the implicit loop, you can use the RETAIN statement.

3.2.1 The RETAIN Statement

Suppose you would like to create a new variable based on values from previous observations, such as creating a variable that accumulates the values from other numeric variables. Consider the following SAS data set, SAS3_1. Based on SAS3_1, suppose you would like to create a new variable, TOTAL, that is used to accumulate the SCORE variable.

SAS3_1:

	ID	SCORE
1	A01	3
2	A02	.
3	A03	4

In order to create an accumulator variable, TOTAL, you need to initialize TOTAL to 0 at the first iteration of the execution. Then at each successive iteration of the execution, add the value from the SCORE variable to the TOTAL variable. Because TOTAL is a new variable that you want to create, TOTAL will be set to missing in the PDV at the beginning of every iteration of the execution. Thus, in order to accumulate the TOTAL variable, you need to retain the value of TOTAL at the beginning of each iteration of the execution. In this situation, you need to use the RETAIN statement. The RETAIN statement has the following form:

RETAIN variable <value>;

In the RETAIN statement, *variable* is the name of the variable that you will want to retain, and *value* is a numeric value that is used to initialize the *variable* only at the first iteration of the DATA step execution. If you do not specify an initial value, the retained variable is initialized as missing before the first execution of the DATA step. The RETAIN statement prevents the *variable* from being initialized each time the DATA step executes. Program 3.3 uses the RETAIN statement to create the TOTAL variable.

The RETAIN statement is a declarative statement and does not execute during the DATA step execution phase. Where you place the RETAIN statement within the DATA step will not affect the value assigned to TOTAL, but where the statement appears *will* affect the column position of the TOTAL variable. For example, placing the RETAIN statement after the SET statement will generate an output data set starting with ID and SCORE and followed by TOTAL because the compiler encounters ID and SCORE in the SET statement before it gets to TOTAL in the RETAIN statement. If you want the resulting data set to begin with the TOTAL variable, you need to place the RETAIN statement, which contains the TOTAL variable, before the SET statement.

Program 3.3:

```
data ex3_2;
    set sas3_1;
    retain total 0;
    total = sum(total, score);
run;

title 'Creating TOTAL by accumulating SCORE';
proc print data = ex3_2;
run;
```

Output from Program 3.3:

```
            Creating TOTAL by accumulating SCORE
            Obs        ID      score     total
             1        A01       3         3
             2        A02       .         3
             3        A03       4         7
```

The execution phase begins immediately after the completion of the compilation phase. At the beginning of the execution phase, the variables ID and SCORE are set to missing (see Figure 3.5); however, the variable TOTAL is initialized to 0 because of the RETAIN statement. Next, the SET statement copies the first observation from the data set SAS3_1 to the PDV. The RETAIN statement is a compile-time only statement; it does not execute during the execution phase. Only then will the variable TOTAL be calculated. Finally, the DATA step execution reaches the final step. Because there is no explicit OUTPUT statement in this program, the implicit OUTPUT statement at the end of the DATA step tells the SAS system to write observations to the data set. The SAS system returns to the beginning of the DATA step to begin the second iteration.

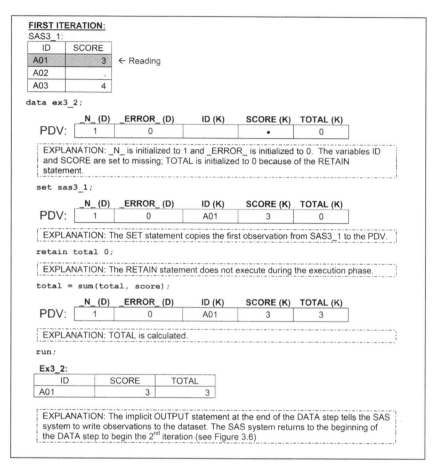

FIGURE 3.5
The first iteration of Program 3.3.

At the beginning of the second iteration, because data is read from an existing SAS data set, instead of reading from the raw data set, values in the PDV for variables ID and SCORE are retained from the previous iteration. The newly created variable TOTAL is also retained because the RETAIN statement is used. See Figures 3.6 and 3.7 for the processes of the second and third iterations. (*Program 3.3* is illustrated in *program3.3.pdf*.)

3.2.2 The SUM Statement

Program 3.3 can be rewritten by using the SUM statement instead of using the RETAIN statement. The SUM statement has the following form:

```
variable+expression;
```

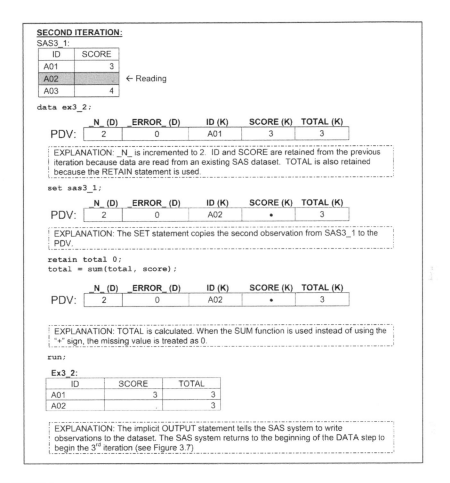

FIGURE 3.6
The second iteration of Program 3.3.

The SUM statement may seem unusual because it does not contain the equal sign. In the SUM statement, *variable* is the numeric accumulator variable that is to be created and is automatically set to 0 at the beginning of the first iteration of the DATA step execution (and is thus retained in following iterations). *Expression* is any SAS expression. In a situation where *expression* is evaluated to a missing value, it is treated as 0. Program 3.4 is an equivalent version of Program 3.3 using the SUM statement.

Program 3.4:

```
data ex3_3;
    set sas3_1;
    total + score;
run;
```

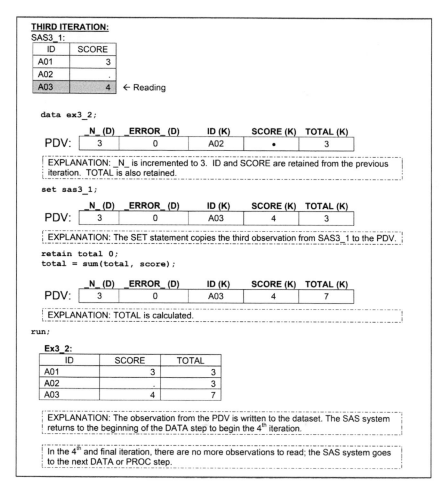

FIGURE 3.7
The third iteration of Program 3.3.

3.3 Conditional Processing in the DATA Step

Many applications require the DATA step to process only part of the observations that meet the condition of a specified expression. In this situation, you need to use the subsetting IF statement.

3.3.1 The Subsetting IF Statement

The subsetting IF statement has the following form:

```
IF expression;
```

The *expression* associated with the subsetting IF statement can be any valid SAS expression. If the *expression* is true for the observation, SAS continues to execute statements in the DATA step and includes the current observation in the data set. The resulting SAS data set contains a subset of the external file or SAS data set. On the other hand, if the *expression* is false, then no further statements are processed for that observation and SAS immediately returns to the beginning of the DATA step. That is to say, the remaining program statements in the DATA step are not executed and the current observation is not written to the output data set.

For example, Program 3.5 creates a data set that contains only the observations in which the SCORE variable is not missing.

Program 3.5:

```
data ex3_4;
    set sas3_1;
    total + score;
    if not missing(score);
run;

title 'Keep observations only when SCORE is not missing';
proc print data = ex3_4;
run;
```

Output from Program 3.5:

```
      Keeping observations for SCORE is not missing
          Obs        ID        score       total
           1         A01         3           3
           2         A03         4           7
```

Sometimes it is more efficient (or easier) to specify a condition for excluding observations instead of including observations in the output data set. In this situation, you can combine the subsetting IF statement with the DELETE statement.

```
IF expression THEN DELETE;
```

Program 3.6 provides an alternative version of Program 3.5 by utilizing the subsetting IF and DELETE statements.

Program 3.6:

```
data ex3_5;
    set sas3_1;
    total + score;
    if missing(score) then delete;
run;
```

3.3.2 Detecting the End of a Data Set by Using the END= Option

Sometimes you may want to create a data set that contains only the contents from the PDV when reading the last observation from the input data set. You can create a temporary variable by using the END= option in the SET statement as a flag to signal when the last observation is being read. The END= option has the following form:

```
SET SAS-data-set END= variable;
```

The *variable* after the keyword END= is a temporary variable that contains an end-of-file indicator. The *variable* is initialized to zero at the beginning of the DATA step iteration and is set to 1 when the SET statement reads the last observation of the input data set. Since the *variable* is a temporary variable, it is not added to the output data set.

Program 3.7 calculates the total score and lists the total number of observations from the data set Sas3_1.

Program 3.7:

```
data total_score(keep = total n);
    set sas3_1 end = last;
    total + score;
    n + 1;
    if last;
run;

title 'Only keep the last observation';
proc print data = total_score;
run;
```

Output from Program 3.7:

```
              Only keep the last observation
                    Obs         total       n
                     1            7          3
```

3.3.3 Restructuring Data Sets from Wide Format to Long Format

Restructuring data sets denotes transforming data from one observation per subject (the *wide* format) to multiple observations per subject (the *long* format) or transforming data from the *long* format to data in the *wide* format. The purpose of the transformation to different formats is to suit the data format requirement for different types of statistical procedures. This type of data transformation can be easily done by using more advanced programming techniques, such as ARRAY processing described in Chapter 6 or the TRANSPOSE procedure described in Chapter 10. However, this can also be accomplished without advanced techniques for simple cases. The example in this chapter is adapted from *Applied Statistics and the SAS® Programming Language* (Cody and Smith, 1991).

Suppose that you are transforming data from the *wide* format (data set WIDE) to the *long* format (data set LONG). Notice that data in the *long* format has a variable, TIME, that distinguishes the different measurements for each subject in the *wide* format. The original variables in the *wide* format, S1–S3, become the variable SCORE in the *long* format.

WIDE:

	ID	S1	S2	S3
1	A01	3	4	5
2	A02	4	.	2

LONG:

	ID	TIME	SCORE
1	A01	1	3
2	A01	2	4
3	A01	3	5
4	A02	1	4
5	A02	3	2

Since only two observations need to be read from the WIDE data set, there will be only two iterations for the DATA step processing. That means you need to generate the output up to three times for each iteration. In some iterations, the output might not be generated three times because missing values in variables S1–S3 will not be output in the LONG data set. Program 3.8 illustrates the data transformation by using multiple OUTPUT statements in one DATA step.

Program 3.8:

```
data long(drop = s1-s3);
    set wide;
    time = 1;
    score = s1;
    if not missing(score) then output;/*OUTPUT # 1*/
    time = 2;
    score = s2;
    if not missing(score) then output;/*OUTPUT # 2*/
    time = 3;
    score = s3;
    if not missing(score) then output;/*OUTPUT # 3*/
run;
```

In Program 3.8, immediately after the SET statement, the TIME variable is set to 1. Next, the value from S1 is assigned to the SCORE variable. Now all the elements for the first observation in the LONG data set are ready for outputting. Before outputting, check whether the SCORE value is missing or not; if it is not missing, use the explicit OUTPUT statement to create the first observation for the LONG data set. Next, assign value 2 to the TIME variable and assign the value from S2 to SCORE and output the data set again. Similar processes are repeated; assign 3 to TIME, assign S3 to SCORE, and then output. Within the first iteration of the DATA step processing, the values for TIME and SCORE are being replaced three times. Once they are replaced, they are output to the final data set. See Figure 3.8 for more details. Note that the automatic variable _ERROR_ is not shown in the figure for simplicity purposes.

The second iteration (see Figure 3.9) is similar to the first iteration. The only difference is that S2 is missing. After the value for S2 is assigned to SCORE, the contents in the PDV are not copied to the final data set because SCORE equals missing. (*Program 3.8* is illustrated in *program3.8.pdf.*)

3.4 Debugging Techniques

Programming errors can be categorized as either syntax or logic errors. Syntax errors are often easier to detect than logic errors since SAS not only stops programs due to syntax errors but also generates detailed error messages in the log window. On the other hand, logic errors often result in generating an unintended data set and they are difficult to debug. This section covers the two most useful debugging strategies for detecting logic errors: utilizing the PUT statement in the DATA step and using the DATA step debugger.

3.4.1 Using the PUT Statement to Observe the Contents of the PDV

A logic error is often created due to a lack of knowing the contents of the PDV during the DATA step execution. If you are not sure what the contents of the PDV are during each step of the DATA step processing, use a PUT statement inside the DATA step, which will generate the contents of each variable in the PDV on the SAS log. This strategy was introduced in the *Longitudinal Data and SAS® A Programmer's Guide* (Cody, 2005). The PUT statement has the following form:

```
PUT variable | variable-list | character-string;
```

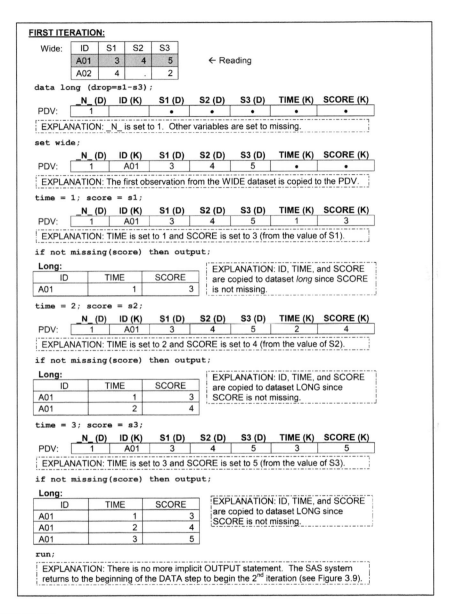

FIGURE 3.8

First iteration for Program 3.8.

The PUT statement can combine explanatory text strings in quotes with the values for selected variables and write the compound output to the SAS log. The keyword _ALL_ (which is an example of the variable-list; see Chapter 1 for details) means that all the variables, including the automatic variables, will be output to the SAS log. For example, Program 3.9 uses the PUT

SECOND ITERATION:

Wide:

ID	S1	S2	S3
A01	3	4	5
A02	4	.	2

← Reading

```
data long (drop=s1-s3);
```

PDV:

N (D)	ID (K)	S1 (D)	S2 (D)	S3 (D)	TIME (K)	SCORE (K)
2	A01	3	4	5	.	.

EXPLANATION: _N_ is set to 2. ID and S1-S3 are retained from the previous iteration. The newly-created variables, TIME and SCORE, are set to missing.

```
set wide;
```

PDV:

N (D)	ID (K)	S1 (D)	S2 (D)	S3 (D)	TIME (K)	SCORE (K)
2	A02	4	.	2	.	.

EXPLANATION: The second observation from WIDE is copied to the PDV.

```
time = 1; score = s1;
```

PDV:

N (D)	ID (K)	S1 (D)	S2 (D)	S3 (D)	TIME (K)	SCORE (K)
2	A02	4	.	2	1	4

EXPLANATION: TIME is set to 1 and SCORE is set to 4 (from the value of S1).

```
if not missing(score) then output;
```

Long:

ID	TIME	SCORE
A01	1	3
A01	2	4
A01	3	5
A02	1	4

EXPLANATION: ID, TIME, and SCORE are copied to data set LONG since SCORE is not missing.

```
time = 2; score = s2;
```

PDV:

N (D)	ID (K)	S1 (D)	S2 (D)	S3 (D)	TIME (K)	SCORE (K)
2	A02	4	.	2	2	.

EXPLANATION: TIME is set to 2 and SCORE is set to missing (from the value of S2).

```
if not missing(score) then output;
```

EXPLANATION: No output is generated since SCORE equals missing.

```
time = 3; score = s3;
```

PDV:

N (D)	ID (K)	S1 (D)	S2 (D)	S3 (D)	TIME (K)	SCORE (K)
2	A02	4	.	2	3	2

EXPLANATION: TIME is set to 3 and SCORE is set to 2 (from the value of S3).

```
if not missing(score) then output;
```

Long:

ID	TIME	SCORE
A01	1	3
A01	2	4
A01	3	5
A02	1	4
A02	3	2

EXPLANATION: ID, TIME, and SCORE are copied to data set LONG since SCORE is not missing.

```
run;
```

EXPLANATION: The SAS system returns to the beginning of the DATA step to begin the 3rd iteration. With no more observations to read in the 3rd iteration, SAS goes to the next DATA or PROC step.

FIGURE 3.9
Second iteration for Program 3.8.

statement to send the contents in the PDV to the SAS log after each statement is executed in the DATA step.

Program 3.9:

```
data ex3_4;
    put "1st PUT" _all_;
    set sas3_1;
    put "2nd PUT" _all_;
    total + score;
    put "3rd PUT" _all_;
    if not missing(score);
    put "4th PUT" _all_;
run;
```

Log from Program 3.9:

```
68 data ex3_4;
69     put "1st PUT" _all_;
70     set sas3_1;
71     put "2nd PUT" _all_;
72     total + score;
73     put "3rd PUT" _all_;
74     if not missing(score);
75     put "4th PUT" _all_;
76 run;


1st PUTID = SCORE =. total = 0 _ERROR_ = 0 _N_ = 1
2nd PUTID = A01 SCORE = 3 total = 0 _ERROR_ = 0 _N_ = 1
3rd PUTID = A01 SCORE = 3 total = 3 _ERROR_ = 0 _N_ = 1
4th PUTID = A01 SCORE = 3 total = 3 _ERROR_ = 0 _N_ = 1
1st PUTID = A01 SCORE = 3 total = 3 _ERROR_ = 0 _N_ = 2
2nd PUTID = A02 SCORE =. total = 3 _ERROR_ = 0 _N_ = 2
3rd PUTID = A02 SCORE =. total = 3 _ERROR_ = 0 _N_ = 2
1st PUTID = A02 SCORE =. total = 3 _ERROR_ = 0 _N_ = 3
2nd PUTID = A03 SCORE = 4 total = 3 _ERROR_ = 0 _N_ = 3
3rd PUTID = A03 SCORE = 4 total = 7 _ERROR_ = 0 _N_ = 3
4th PUTID = A03 SCORE = 4 total = 7 _ERROR_ = 0 _N_ = 3
1st PUTID = A03 SCORE = 4 total = 7 _ERROR_ = 0 _N_ = 4
NOTE: There were 3 observations read from the data set WORK.
SAS3_1.
NOTE: The data set WORK.EX3_4 has 2 observations and 3
variables.
NOTE: DATA statement used (Total process time):
      real time      0.01 seconds
      cpu time       0.03 seconds
```

3.4.2 Using the DATA Step Debugger

SAS DATA step debugger is a part of Base SAS software and is described in the "DATA Step Debugger" article of SAS documentation.

You can use the DATA step debugger to identify logic errors by examining the contents of the PDV interactively while executing one DATA step at a time. The DATA step debugger cannot be used for a PROC step. To invoke the DATA step debugger, you need to add the DEBUG option to the DATA statement. Program 3.10 illustrates how to use the DATA step debugger.

Program 3.10:

```
data ex3_4/debug;
    set sas3_1;
    total + score;
    if not missing(score);
run;
```

Submitting a DATA step that contains the DEBUG option will invoke the DEBUGGER LOG and DEBUGGER SOURCE windows. You can enter the debugger commands after the prompt (>) in the last line of the DEBUGGER LOG. All the generated results from the debugger commands will be recorded in the DEBUGGER LOG window. DATA step debugger contains numerous commands that allow you to execute, bypass, or suspend one or more statements, examines the values of selected variables in the PDV at any point of execution, displays the attributes of selected variables, and so on. Only a small number of commands are introduced in this section.

In the DEBUGGER SOURCE window, the line numbers that correspond to the program are the same as the ones in the SAS log window. The DEBUGGER SOURCE window enables you to view the position in the DATA step while you debug your program. The DATA step execution pauses before the execution of the statement that is being highlighted. Notice that the current highlighted statement is the SET statement, which is immediately the next statement to be executed. At this point, to display the contents in the PDV, you can use the EXAMINE debugger command, which has the following form:

EXAMINE _ALL_ <format> | variable-1 <format-1> <...variable-n
          <format-n>>

In the EXAMINE command, you can either use the keyword _ALL_ to display the contents of all variables or only the selected variables in the PDV. The *format* option can be used to display the variables in either SAS built-in or user-defined formats. For example, to display the contents of all variables in the PDV, you can submit the following command:

```
examine _all_
```

Here are the generated results from the submitted statement above:

Debugger Log:

```
Stopped at line 30 column 5
> examine _all_
ID =
score = .
total = 0
_ERROR_ = 0
_N_ = 1
```

If you want to examine the content for only one variable, for example, TOTAL, you can type the following command line:

```
examine total
```

In the debugging mode, you can execute the DATA step either one statement or several statements at a time by using the STEP command. The general form of the STEP command is as follows:

STEP <n>

The optional *n* in the STEP command is used to specify the number of statements to execute. If you want to execute only one statement at the moment, you can simply hit the ENTER key because by default the STEP command is associated with the ENTER key.

Another useful debugging command is the BREAK command, which is used to suspend the execution of a program at an executable statement. The BREAK command has the following form:

BREAK location <WHEN expression>

You can specify a line number at the *location* at which to set a breakpoint. The BREAK command is often used with the GO command. The GO command is used to start or resume the execution of the DATA step. For example, if you want to execute the DATA step until it reaches line 47, you can submit the following two statements:

```
break 47
go
```

In some situations, stopping at a certain line in the DATA step at each itera-tion of the DATA step execution might not be efficient, especially when you are reading a large data set. Thus, you can utilize the WHEN *expression* in the BREAK command to set up a breakpoint when a certain condition is met. For example, the following command set a breakpoint at line 47 when the SCORE variable is missing:

```
break 47 when score = .
```

To stop the debugging mode of the DATA step execution, you can use the QUIT command:

QUIT

Submitting the QUIT command terminates a debugger session and returns control to the SAS session.

Exercises

Exercise 3.1. Consider the following data set, PROB3_1.SAS7BDAT:

	ID	SCORE
1	A	3
2	B	4
3	C	.
4	D	5
5	E	.
6	F	.

Notice that the SCORE variable contains missing values for some observations. For this exercise, you need to modify the SCORE variable. If SCORE is missing for the current observation, use the SCORE value from the previous observation. The resulting data will look as shown below:

	ID	SCORE
1	A	3
2	B	4
3	C	4
4	D	5
5	E	5
6	F	5

Exercise 3.2. Consider the following data set, PROB3_2.SAS7BDAT:

	ID	GROUP	S1	S2	S3
1	A01	A	3	4	5
2	A01	B	2	3	9
3	A02	A	4	.	2
4	A02	B	4	5	3

Restructure PROB3_2.SAS7BDAT to the format exactly like as shown below. The variable MEASURE can be created by using the CATS function, which is used to concatenate character strings. You can find out how to use the CATS function in Chapter 9 or in the "CATS Function" article of SAS documentation.

	ID	MEASURE	SCORE
1	A01	A1	3
2	A01	A2	4
3	A01	A3	5
4	A01	B1	2
5	A01	B2	3
6	A01	B3	9
7	A02	A1	4
8	A02	A3	2
9	A02	B1	4
10	A02	B2	5
11	A02	B3	3

4

BY-Group Processing in the DATA Step

4.1 Introduction to BY-Group Processing

Most of the examples in previous chapters used a data set that contains only one observation per subject. Sometimes you will also work with data with multiple observations per subject. This type of data often results from repeated measures for each subject and is often called longitudinal data. *Longitudinal Data and SAS® A Programmer's Guide* (Cody, 2005) provides many useful applications for manipulating longitudinal data.

Applications that involve longitudinal data often require identifying the beginning or end of measurement for each subject. This can be accomplished by using the BY-group processing method. *BY-group processing* is a method of processing records from data sets that can be grouped by the values of one or more common variables. These "grouping" variables are called the *BY variable*. The value of a BY variable is called the *BY value*. A *BY group* refers to all observations with the same BY value. When there are multiple variables designated as BY variables, a BY group would be a group of records with the same combination of the values of these BY variables with each BY group containing a unique combination of values for the BY variables. During BY-group processing, SAS creates two automatic variables, FIRST.VARIABLE and LAST.VARIABLE, which are used to indicate the beginning or the end of the measurement within each BY group.

4.1.1 The FIRST.VARIABLE and the LAST.VARIABLE

In order to utilize BY-group processing, you need to place a BY statement with one or more BY variable(s) after the SET statement. Furthermore, the input data set also needs to be previously sorted by the BY variable(s).

During BY-group processing, SAS creates two temporary indicator variables for each BY variable: FIRST.VARIABLE and LAST.VARIABLE. Because FIRST.VARIABLE and LAST.VARIABLE are temporary variables, they are not sent to the output data set. Both FIRST.VARIABLE and LAST.VARIABLE are initialized to 1 at the beginning of the DATA step execution. Then FIRST.VARIABLE is set to 1 in the program data vector (PDV) when SAS reads the first observation in each BY group and is set to 0 when reading the second to the last observation in each BY group. Similarly, LAST.VARIABLE is set to 1 when reading the last observation in each BY group and set to 0 when reading those observations that are not last.

Consider the following SAS data set, SAS4_1, which consists of five observations with the values of SCORE for two subjects, A01 and A02.

SAS4_1:

	ID	SCORE
1	A01	3
2	A01	3
3	A01	2
4	A02	4
5	A02	2

Suppose that the ID variable is the representative BY variable; consequently, there will be two BY groups because there are two distinct values for the ID variable. FIRST.ID and LAST.ID will be created and represented as in Figure 4.1. If you use ID and SCORE as the BY variables, then in addition to FIRST.ID and LAST.ID, FIRST.SCORE and LAST.SCORE will be created in

	ID	SCORE	GROUP: ID	FIRST.ID	LAST.ID
1	A01	3		1	0
2	A01	3	1	0	0
3	A01	2		0	1
4	A02	2	2	1	0
5	A02	4		0	1

	ID	SCORE	GROUP: ID & SCORE	FIRST.SCORE	LAST.SCORE
1	A01	3	1	1	0
2	A01	3		0	1
3	A01	2	2	1	1
4	A02	2	3	1	1
5	A02	4	4	1	1

FIGURE 4.1
The top table illustrates the values for automatic variables FIRST.ID and LAST.ID. The bottom table illustrates the values for FIRST.SCORE and LAST.SCORE.

the PDV. There will be four BY groups based on unique combination values of ID and SCORE.

4.1.2 The Execution Phase of BY-Group Processing

Suppose that you would like to calculate the total scores for each subject in the SAS4_1 data set. To create a variable (TOTAL) that is the total score for each subject, you need to initialize TOTAL to 0 when starting to read the first observation of each subject. Then TOTAL can be accumulated by adding the value from the SCORE variable to TOTAL for each observation. In the end, you can output the TOTAL score when reading the last observation of each subject. Therefore, you need to utilize BY-group processing and use ID as the BY variable. The solution for this problem is shown in Program 4.1.

Program 4.1:

```
proc sort data = sas4_1;
    by id;
run;
data sas4_2 (drop = score);
    set sas4_1;
    by id;
    if first.id then total = 0;
    total + score;/* SUM statement */
    if last.id;
run;

title 'The total scores for each subject';
proc print data = sas4_2;
run;
```

Output from Program 4.1:

```
        The total scores for each subject
            Obs    ID     total
             1     A01       8
             2     A02       6
```

Because the BY statement with ID as the BY variable was used after the SET statement in the DATA step, two automatic variables, FIRST.ID and LAST.ID, are created in the PDV. Both FIRST.ID and LAST.ID are initialized to 1 before the first iteration of the DATA step execution (see Figure 4.2). ID and SCORE variables are set to missing, but TOTAL is set to 0 because TOTAL is created by the SUM statement. (The SUM statement is the one after the IF-THEN statement.) When the SET statement executes, SAS copies the first observation from the sorted SAS4_1 data set to the PDV. Because this is the first observation for the A01 subject, FIRST.ID is set to 1. The LAST.ID is

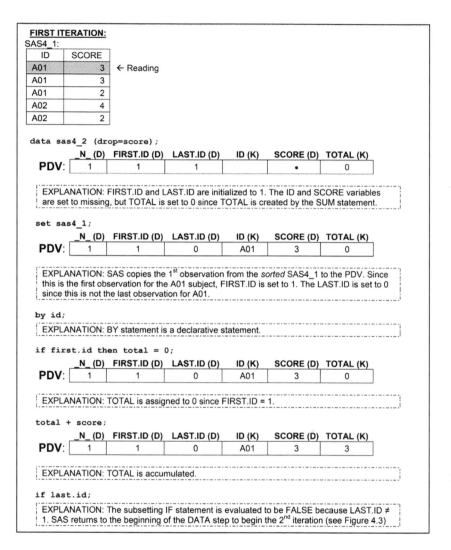

FIGURE 4.2
First iteration for Program 4.1.

set to 0 because this is not the last observation. Next, TOTAL is assigned to 0 because this is the first observation for ID A01 (SAS statement: if first.id then total = 0;). Then, the SUM statement accumulates the TOTAL variable. Because the subsetting IF statement is evaluated to be false (LAST.ID does not equal 1), SAS immediately returns to the beginning of the DATA step. That means the contents of the PDV are not sent to the output data set, SAS4_2.

The second iteration (see Figure 4.3) is similar to the first iteration. During this iteration, both FIRST.ID and LAST.ID are set to 0. TOTAL is then

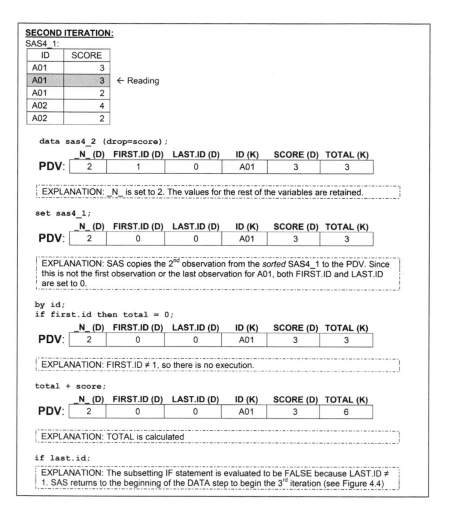

FIGURE 4.3
Second iteration for Program 4.1.

accumulated. But again the PDV contents are not output to the SAS data set because this is not the last observation for A01.

In the third iteration (see Figure 4.4), FIRST.ID is set to 0 and LAST.ID is set to 1. TOTAL is accumulated. The subsetting IF statement is evaluated to be true. Then SAS reaches the end of the DATA step and the implicit OUTPUT statement copies the contents from the PDV (ID and TOTAL) to the SAS data set sas4_2 (the SCORE variable is dropped in the DATA statement). The rest of the iterations are similar to the iterations explained above. See Figures 4.5 and 4.6 for details. (*Program 4.1* is illustrated in *program4.1.pdf.*)

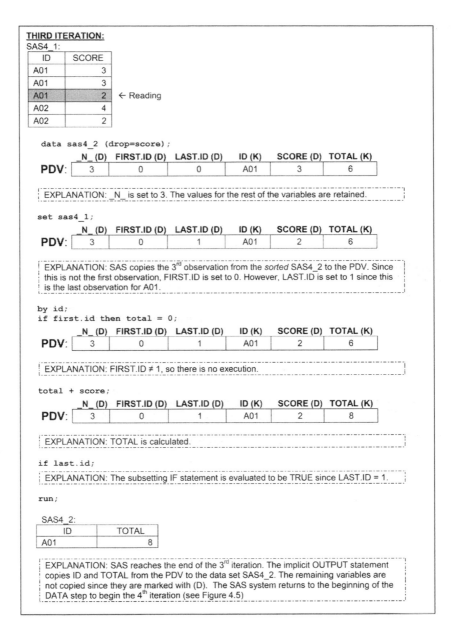

FIGURE 4.4

Third iteration for Program 4.1.

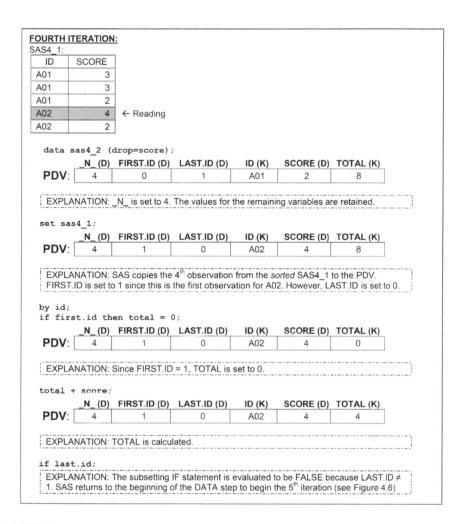

FOURTH ITERATION:
SAS4_1:

ID	SCORE
A01	3
A01	3
A01	2
A02	4	← Reading
A02	2

```
data sas4_2 (drop=score);
```

	_N_ (D)	FIRST.ID (D)	LAST.ID (D)	ID (K)	SCORE (D)	TOTAL (K)
PDV:	4	0	1	A01	2	8

EXPLANATION: _N_ is set to 4. The values for the remaining variables are retained.

```
set sas4_1;
```

	_N_ (D)	FIRST.ID (D)	LAST.ID (D)	ID (K)	SCORE (D)	TOTAL (K)
PDV:	4	1	0	A02	4	8

EXPLANATION: SAS copies the 4th observation from the *sorted* SAS4_1 to the PDV. FIRST.ID is set to 1 since this is the first observation for A02. However, LAST.ID is set to 0.

```
by id;
if first.id then total = 0;
```

	_N_ (D)	FIRST.ID (D)	LAST.ID (D)	ID (K)	SCORE (D)	TOTAL (K)
PDV:	4	1	0	A02	4	0

EXPLANATION: Since FIRST.ID = 1, TOTAL is set to 0.

```
total + score;
```

	_N_ (D)	FIRST.ID (D)	LAST.ID (D)	ID (K)	SCORE (D)	TOTAL (K)
PDV:	4	1	0	A02	4	4

EXPLANATION: TOTAL is calculated.

```
if last.id;
```
EXPLANATION: The subsetting IF statement is evaluated to be FALSE because LAST.ID ≠ 1. SAS returns to the beginning of the DATA step to begin the 5th iteration (see Figure 4.6)

FIGURE 4.5
Fourth iteration for Program 4.1.

4.2 Applications Utilizing BY-Group Processing

Longitudinal data does not always require BY-group processing when it is read into a SAS data set. Instead, a program that uses BY-group processing does so to identify either the beginning or the end of a measurement within a BY group. Therefore, a DATA step that uses BY-group processing frequently contains the following:

1. A cumulating variable is initialized to 0 when the FIRST.VARIABLE equals 1.

FIFTH ITERATION:

SAS4_1:

ID	SCORE
A01	3
A01	3
A01	2
A02	4
A02	2	← Reading

```
data sas4_2 (drop=score);
```

	_N_ (D)	FIRST.ID (D)	LAST.ID (D)	ID (K)	SCORE (D)	TOTAL (K)
PDV:	5	1	0	A02	4	4

EXPLANATION: _N_ is set to 5. The values for the remaining variables are retained.

```
set sas4_1;
```

	_N_ (D)	FIRST.ID (D)	LAST.ID (D)	ID (K)	SCORE (D)	TOTAL (K)
PDV:	5	0	1	A02	2	4

EXPLANATION: SAS copies the 5th observation from the *sorted* SAS4_1 to the PDV. FIRST.ID is set to 0. LAST.ID is set to 1 since this is the last observation for A02.

```
by id;
if first.id then total = 0;
```

	_N_ (D)	FIRST.ID (D)	LAST.ID (D)	ID (K)	SCORE (D)	TOTAL (K)
PDV:	5	0	1	A02	2	4

EXPLANATION: FIRST.ID $\neq$ 1, so there is no execution.

```
total + score;
```

	_N_ (D)	FIRST.ID (D)	LAST.ID (D)	ID (K)	SCORE (D)	TOTAL (K)
PDV:	5	0	1	A02	2	6

EXPLANATION: TOTAL is calculated.

```
if last.id;
```

EXPLANATION: The subsetting IF statement is evaluated to be TRUE since LAST.ID = 1.

```
run;
```

SAS4_2:

ID	TOTAL
A01	8
A02	6

EXPLANATION: SAS reaches the end of the 5th iteration. The implicit OUTPUT statement copies ID and TOTAL from the PDV to the data set SAS4_2. The SAS system returns to the beginning of the DATA step to begin the 6th iteration. Since there are no more observations to read in the 6th iteration, the SAS system goes to the next DATA or PROC step.

FIGURE 4.6
Fifth iteration for Program 4.1.

2. A cumulating variable is accumulated with some values at every iteration of the DATA step.

3. Some calculation needs to be performed when the LAST.VARIABLE equals 1.

4. The contents of the PDV are output only when the LAST.VARIABLE equals 1.

5. In addition to the BY variable, another variable will need to be previously sorted. However, only the BY variable is used in the SET statement in the DATA step.

Most applications don't complete all five steps in the list above. For example, Program 4.1 completes only three:

- TOTAL is initialized to 0 when FIRST.ID equals 1 (# 1).
- TOTAL accumulates its value from the SCORE variable at every iteration of the DATA step (# 2).
- At the end of the DATA step, the contents from the PDV are written to the output data set when LAST.ID equals 1 (# 4).

This section covers some commonly encountered applications that utilize BY-group processing.

4.2.1 Calculating Mean Score within Each BY Group

Suppose that you would like to calculate the mean score for each person based on the data set SAS4_1. One way to accomplish this task is to use the MEANS procedure and use SCORE in the VAR statement and ID in the CLASS statement. Alternatively, you can also use BY-group processing in the DATA step to create a data set that contains the mean score for each subject.

The solution for this problem is similar to the one in Program 4.1. Even though you are calculating the means instead of the total score, you still need to accumulate all the scores for each subject (TOTAL) and create a "counter" variable (N) to count the number of observations within each BY group. Furthermore, TOTAL and N need to be initialized to 0 when FIRST. ID equals 1. Then you will need to calculate the mean score (MEAN_SCORE) by dividing TOTAL by N and output the result when LAST.ID equals 1. (See Program 4.2.)

Program 4.2:

```
data sas4_mean (drop = score);
    set sas4_1;
    by id;
```

```
   if first.id then do;
       total = 0;
       n = 0;
   end;
   total + score;
   n + 1;
   if last.id then do;
       mean_score = total/n;
       output;
   end;
run;

title 'The mean score for each subject';
proc print data = sas4_mean;
run;
```

Output from Program 4.2:

```
    The mean score for each subject
                               mean_
    Obs     ID      total   n   score
     1      A01       8     3   2.66667
     2      A02       6     2   3.00000
```

4.2.2 Creating Data Sets with Duplicate or Non-Duplicate Observations

A common task in examining a data set is checking when the data set contains duplicate observations. Again, you can use BY-group processing to identify duplicated observations.

The first two observations in SAS4_1 are identical. Suppose that you would like to create two data sets: one with observations with non-duplicated records and one containing observations with duplicated records from the data set SAS4_1. Because a duplicated record will have the same value for both the ID and SCORE variables, both ID and SCORE variables will be used as the BY variables. A non-duplicated record is the one where both FIRST. SCORE and LAST.SCORE equal 1; otherwise, it will be a duplicated record. The solution for this program is in Program 4.3.

Program 4.3:

```
proc sort data = sas4_1;
    by id score;
run;

data sas4_1_s sas4_1_d;
    set sas4_1;
    by id score;
    if first.score and last.score then output sas4_1_s;
```

```
      else output sas4_1_d;
run;

title 'Non-duplicated records';
proc print data = sas4_1_s;
run;

title 'Duplicated records';
proc print data = sas4_1_d;
run;
```

Output from Program 4.3:

```
              Non-duplicated records
              Obs     ID     SCORE
               1      A01      2
               2      A02      2
               3      A02      4

              Duplicated records
              Obs     ID     SCORE
               1      A01      3
               2      A01      3
```

4.2.3 Obtaining the Most Recent Non-Missing Data within Each BY Group

Longitudinal data, such as patients with repeated measurements over time, are often encountered in the clinical field. For example, a patient may have multiple measurements of weight, blood pressure, total cholesterol, or blood glucose level over several medical visits, but may not have all these values recorded every time. Researchers may be interested in a database depicting the most recent available information on their patients.

For example, the data set PATIENT contains the triglyceride (TGL) measurement and smoking status (SMOKE) for patients for different time periods. Notice that some patients have only one measurement, whereas others were measured more than once in different years. Suppose that you would like to create a data set that contains the most recent non-missing data. The resulting data set will have three variables: PATID (patient ID), TGL_NEW (the most recent non-missing TGL), and SMOKE_NEW (the most recent non-missing SMOKE). A couple of issues need to be considered for solving this problem. First, the most recent non-missing data for TGL and SMOKE occur at different time points. For instance, for patient A01, the most recent non-missing TGL is 150 in 2007 but the most recent non-missing SMOKE is "Y" in 2005. The second issue is that some patients might have missing values for either TGL or SMOKE. In this situation, you need to use the missing variable in the resulting data set for this variable. For instance, the TGL measurement is missing for A03.

PATIENTS:

	PATID	VISIT	TGL	SMOKE
1	A01	2005	.	Y
2	A01	2007	150	
3	A02	2004	.	
4	A02	2005	200	N
5	A02	2006	210	N
6	A03	2005	.	Y
7	A04	2002	164	
8	A04	2004	170	Y
9	A04	2006	190	
10	A04	2007	.	N
11	A05	2005	189	

Because you need to keep only the most recent record, the data set has to be sorted by the PATID and VISIT variables in ascending order. However, when utilizing BY-group processing in the DATA step, you need to use PATID as the BY variable. One idea for solving this problem is that you initially assign TGL_NEW and SMOKE_NEW to missing values when reading the first observation for each patient from the sorted data set. At each iteration of the DATA step, you will assign the values from the TGL and SMOKE variables to TGL_NEW and SMOKE_NEW, respectively, provided that TGL and SMOKE are not missing. Then you will output the values in the PDV when reading the last observation of each patient. Because TGL_NEW and SMOKE_NEW are newly created variables, you need to retain their values by using the RETAIN statement. The solution for this problem is illustrated in Program 4.4.

Program 4.4:

```
proc sort data = patients out = patients_sort;
    by patid visit;
run;

data patients_single (drop = visit tgl smoke);
    set patients_sort;
    by patid;
    retain tgl_new smoke_new;
    if first.patid then do;
        tgl_new =.;
        smoke_new = " ";
    end;
    if not missing(tgl) then tgl_new = tgl;
    if not missing(smoke) then smoke_new = smoke;
    if last.patid;
run;
```

```
title 'The data contains the most current non-missing values';
proc print data = patients_single;
run;
```

Output from Program 4.4:

```
      The data contains the most current non-missing values
                                                smoke_
           Obs      patid      tgl_new          new
            1        A01         150             Y
            2        A02         210             N
            3        A03          .              Y
            4        A04         190             N
            5        A05         189
```

4.2.4 Restructuring Data Sets from Long Format to Wide Format

Restructuring data sets from the *wide* format to the *long* format was illustrated in Chapter 3. You can also transform the data set from the *long* format to the *wide* format by using BY-group processing. A solution to solve this problem is illustrated in Program 4.5. A more efficient way to solve this problem is shown in Chapter 6, "Array Processing." Data set LONG from Chapter 3 is used again as input for Program 4.5. This time, WIDE is generated as output.

Program 4.5:

```
proc sort data = long;
    by id time;
run;

data wide (drop = time score);
    set long;
    by id;
    retain s1-s3;
    if first.id then do;
        s1 =.; s2 =.; s3 =.;
    end;
    if time = 1 then s1 = score;
    else if time = 2 then s2 = score;
    else s3 = score;
    if last.id;
run;
```

Program 4.5 begins by sorting the LONG data set by ID and TIME. Sorting the variable TIME within each ID is important because it ensures that the horizontal order of S1–S3 in the WIDE data set for each subject can be matched correctly with the vertical order of SCORE in the LONG data set.

Because you are reading five observations from the LONG data set but creating only two observations, it means that you are *not* copying data from the PDV to the final data set at each iteration. As a matter of fact, you need to generate only one observation once all the observations for each subject have been processed. That means that the newly created variables (S1–S3) in the final data set need to retain their values; otherwise S1–S3 will be initialized to missing at the beginning of each iteration of the DATA step processing.

Notice that subject A02 is missing one observation for TIME equaling 2. The value of S2 from the previous subject (A01) will be copied to the data set WIDE for the subject A02 instead of a missing value because S2 is being retained. To avoid this problem, initialize S1–S3 to missing when processing the first observation for each subject. (*Program 4.5* is illustrated in *program4.5.pdf*.)

Exercises

Exercise 4.1. Consider the following data set, PROB4_1.SAS7BDAT:

	ID	TIME	SCORE
1	A	1	3
2	A	2	.
3	A	3	4
4	B	1	.
5	B	2	5
6	B	3	.

The SCORE variable is recorded at three different time points for each subject. For this exercise, you need to modify the SCORE variable. If SCORE is missing for the current observation, use the SCORE value from the previous recorded time point within each ID.

The resulting data will look as shown below:

	ID	TIME	SCORE
1	A	1	3
2	A	2	3
3	A	3	4
4	B	1	.
5	B	2	5
6	B	3	5

Exercise 4.2. Program 4.4 creates a data set that contains the most recent non-missing data within each BY group. Modify this program by adding two additional variables, TGL_YEAR and SMOKE_YEAR, that contain the corresponding year when the most recent non-missing data were measured. The resulting data will look like as shown below:

	PATID	TGL_NEW	SMOKE_NEW	TGL_YEAR	SMOKE_YEAR
1	A01	150	Y	2007	2005
2	A02	210	N	2006	2006
3	A03	.	Y	.	2005
4	A04	190	N	2006	2007
5	A05	189		2005	.

Exercise 4.3. Use the same data set from Exercise 4.2 (PATIENTS. SAS7BDAT) to create a new SAS data set that contains only the first two visits of each patient. If a patient has only one visit, you will need to keep only this one observation for this patient. The resulting data will look as shown below:

	PATID	VISIT	TGL	SMOKE
1	A01	2005	.	Y
2	A01	2007	150	
3	A02	2004	.	
4	A02	2005	200	N
5	A03	2005	.	Y
6	A04	2002	164	
7	A04	2004	170	Y
8	A05	2005	189	

5

Writing Loops in the DATA Step

5.1 Implicit and Explicit Loops

A loop is a basic logical programming concept where one or more statements are executed repetitively until a predefined condition is satisfied. Loops are used in all programming languages. Compared to other programming languages, loop processing is more complex in SAS® because of its unique *implicit* loop. The implicit loop is not an easy construct to grasp, even by programmers with advanced skills in other programming languages. Knowing when the time is right to create an *explicit* loop can also present a challenge to the SAS programmer who has to work simultaneously with the implied loop in the DATA step. The material in this chapter is based on a paper that I presented at SAS Global Forum (Li, 2010).

5.1.1 Implicit Loops

An implicit loop, which was introduced in Chapter 3, results when the DATA step repetitively reads data values from the input data set, executes statements, and creates observations for the output data set (one at a time) during the execution phase. SAS stops reading the input file when it reaches the end-of-file marker, which is located at the end of the input data file; at this point, the implicit loop ends.

The following example shows how the implicit loop is processed. Suppose that you would like to assign each subject in a group of patients in a clinical trial where each patient has a 50% chance of receiving either the drug or a placebo. For illustration purposes, only four patients from the trial are used in this example. The data set is similar to the one below. The solution is shown in Program 5.1.

PATIENT:

	ID
1	M2390
2	F2390
3	F2340
4	M1240

Program 5.1:

```
data ex5_1 (drop = rannum);
    set patient;
    rannum = ranuni(2);
    if rannum > 0.5 then group = 'D'; /* DRUG */
    else group = 'P';                  /* PLACEBO */
run;


title 'Assigning patients to different groups';
proc print data = ex5_1;
run;
```

Output from Program 5.1:

```
        Assigning patients to different groups
            Obs        ID        group
             1        M2390         P
             2        F2390         D
             3        F2340         D
             4        M1240         D
```

In Program 5.1, The RANUNI function is used to generate a random number that follows uniform distribution between 0 and 1. The general form has the following:

RANUNI (seed)

The *seed* in the RANUNI function is a nonnegative integer. The RANUNI function generates a stream of numbers based on *seed*. When *seed* is set to 0, which is the computer clock, the generated number cannot be reproduced. However, when *seed* is a non-zero number, the generated number can be reproduced.

The DATA step execution in Program 5.1 consists of four iterations. Within each iteration, SAS reads one observation from the input data set first by using the SET statement. Then a random number is generated and assigned to the RANNUM variable. Next, the GROUP variable is assigned with value 'D' if RANNUM is greater than 0.5; otherwise, GROUP is assigned with value 'P.' At the end of the iteration, an implicit OUTPUT statement tells SAS to write the contents from the PDV to the output data set EX5_1. The entire execution process in Program 5.1 uses an implicit loop.

5.1.2 Explicit Loops

In the previous example, the patient ID is stored in an input data set. Suppose you don't have a data set containing patient IDs. You are asked to assign four patients with a 50% chance of receiving either the drug or the placebo. Instead of creating an input data set that stores ID, you can create the ID and assign each ID to a group in the DATA step at the same time. (See Program 5.2.)

Program 5.2:

```
data ex5_2(drop = rannum);
    id = 'M2390';
    rannum = ranuni(2);
    if rannum > 0.5 then group = 'D';
    else group = 'P';
    output;

    id = 'F2390';
    rannum = ranuni(2);
    if rannum > 0.5 then group = 'D';
    else group = 'P';
    output;

    id = 'F2340';
    rannum = ranuni(2);
    if rannum > 0.5 then group = 'D';
    else group = 'P';
    output;

    id = 'M1240';
    rannum = ranuni(2);
    if rannum > 0.5 then group = 'D';
    else group = 'P';
    output;
run;
```

The DATA step in Program 5.2 begins with assigning ID numbers and then assigns the group value based on a generated random number. There are four explicit OUTPUT statements that tell SAS to write the current observation from the PDV to the SAS data set immediately, not at the end of the DATA step. However, without using the explicit OUTPUT statement, you will only create one observation for ID = M1240. Notice that most of the statements above are identical. To reduce the amount of coding, you can simply rewrite the program by placing repetitive statements in a DO loop. Following is the general form for an iterative DO loop:

```
DO index-variable = value1, value2, ..., valuen;
SAS statements
END;
```

In the iterative DO loop, you must specify an *index-variable* that contains the value of the current iteration. The loop will execute along *value*1 through *valuen* and the *values* can be either character or numeric. Program 5.3 is an improvement over Program 5.2 because it utilizes an iterative DO loop.

Program 5.3:

```
data ex5_3(drop = rannum);
    do id = 'M2390', 'F2390', 'F2340', 'M1240';
        rannum = ranuni(2);
        if rannum > 0.5 then group = 'D';
        else group = 'P';
        output;
    end;
run;
```

More often the iterative DO loop along a sequence of integers is used.

```
DO index-variable = start TO stop <BY increment>;
SAS statements
END;
```

The loop will execute from the *start* value to the *stop* value. The optional BY clause specifies an *increment* between *start* and *stop*. The default value for the *increment* is 1; *start, stop,* and *increment* can be numbers, variables, or SAS expressions. These values are set upon entry into the DO loop and cannot be modified during the processing of the DO loop. However, the *index-variable* can be changed within the loop.

Suppose that you are using a numeric sequence, say 1 to 4, as patient IDs; you can rewrite the previous program as Program 5.4.

Program 5.4:

```
data ex5_4 (drop = rannum);
    do id = 1 to 4;
        rannum = ranuni(2);
        if rannum > 0.5 then group = 'D';
        else group = 'P';
        output;
    end;
run;
```

Program 5.4 didn't specify the *increment* value; the default value of one is assumed. You can also decrement a DO loop by using a negative value, such as –1. The execution phase is illustrated in Figures 5.1 and 5.2.

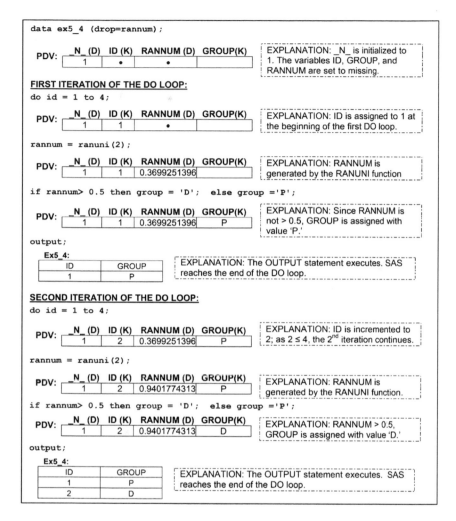

FIGURE 5.1
The first two iterations of the DO loop in Program 5.4.

Program 5.4 uses an iterative DO loop requiring that you specify the *number* of iterations for the DO loop. Sometimes, however, you will need to execute statements repetitively until a condition is met. In this situation, you need to use either the DO WHILE or DO UNTIL statements. Following is the syntax for the DO WHILE statement:

```
DO WHILE (expression);
SAS statements
END;
```

THIRD ITERATION OF THE DO LOOP:
```
do id = 1 to 4;
```

PDV:

N_ (D)	ID (K)	RANNUM (D)	GROUP(K)
1	3	0.9401774313	D

EXPLANATION: ID is incremented to 3; as 3 ≤ 4, the 3rd iteration continues.

```
rannum = ranuni(2);
```

PDV:

N_ (D)	ID (K)	RANNUM (D)	GROUP(K)
1	3	0.7996486122	D

EXPLANATION: RANNUM is generated by the RANUNI function

```
if rannum> 0.5 then group = 'D';  else group ='P';
```

PDV:

N_ (D)	ID (K)	RANNUM (D)	GROUP(K)
1	3	0.7996486122	D

EXPLANATION: RANNUM > 0.5, GROUP is assigned with value 'D.'

```
output;
```

Ex5_4:

ID	GROUP
1	P
2	D
3	D

EXPLANATION: The OUTPUT statement executes. SAS reaches the end of DO loop.

FOURTH ITERATION OF THE DO LOOP:
```
do id = 1 to 4;
```

PDV:

N_ (D)	ID (K)	RANNUM (D)	GROUP(K)
1	4	0.7996486122	D

EXPLANATION: ID is incremented to 4; as 4 ≤ 4, the 4th iteration continues.

```
rannum = ranuni(2);
```

PDV:

N_ (D)	ID (K)	RANNUM (D)	GROUP(K)
1	4	0.5187972908	D

EXPLANATION: RANNUM is generated by the RANUNI function.

```
if rannum> 0.5 then group = 'D';  else group ='P';
```

PDV:

N_ (D)	ID (K)	RANNUM (D)	GROUP(K)
1	4	0.5187972908	D

EXPLANATION: RANNUM > 0.5, GROUP is assigned with value 'D.'

```
output;
```

Ex5_4:

ID	GROUP
1	P
2	D
3	D
4	D

EXPLANATION: The OUTPUT statement executes. SAS reaches the end of DO loop.

```
end;
```

FIFTH ITERATION OF THE DO LOOP:

PDV:

N_ (D)	ID (K)	RANNUM (D)	GROUP(K)
1	5	0.5187972908	D

EXPLANATION: ID is incremented to 5; since 5 is > 4, the loop ends.

```
run;
```

EXPLANATION: There will be no explicit OUTPUT statement. Since we didn't read an input data set, the DATA step execution ends.

FIGURE 5.2
The last three iterations of the DO loop in Program 5.4.

In the DO WHILE loop, the *expression* is evaluated at the top of the DO loop. The DO loop will not execute if the *expression* is false. We can rewrite *program5.4* by using the DO WHILE loop.

Program 5.5:

```
data ex5_5(drop = rannum);
    do while (id < 4);
        id + 1;
        rannum = ranuni(2);
        if rannum > 0.5 then group = 'D';
        else group = 'P';
        output;
    end;
run;
```

In Program 5.5, the ID variable is created inside the loop with the SUM statement. Thus, the variable ID is initialized to 0 in the PDV at the beginning of the DATA step execution, which precedes the DO WHILE statement. At the beginning of the loop, the condition (ID < 4) is checked. Since ID equals 0, which satisfies the condition, the first iteration begins. Per iteration, the ID variable is accumulated from the SUM statement. Iterations continue until the condition (ID < 4) is not met.

Alternatively, you can also use the DO UNTIL loop to execute the statements conditionally. Unlike DO WHILE loops, the DO UNTIL loop evaluates the condition at the end of the loop. That means the DO UNTIL loop always executes at least once. The DO UNTIL loop follows the form below:

```
DO UNTIL (expression);
SAS statements
END;
```

Program 5.6 is a copy of Program 5.5 except that DO UNTIL replaces DO WHILE, and the condition in *expression* is changed to ID = 4.

Program 5.6:

```
data ex5_6(drop = rannum);
    do until (id = 4);
        id + 1;
        rannum = ranuni(2);
        if rannum > 0.5 then group = 'D';
        else group = 'P';
        output;
    end;
run;
```

A type of programming error that a programmer might encounter is writing infinite loops. An infinite loop causes the SAS statements within the loop to execute endlessly. The main cause of infinite looping is that the condition for stopping the loop is specified incorrectly. One way to terminate an infinite loop is to click on the "Break" button. The "Break" button has the symbol of explanation point on the tool bar.

5.1.3 Nested Loops

You can place a loop within another loop. To continue with the previous example, suppose that you would like to assign 12 subjects from 3 cancer centers ("COH," "UCLA," and "USC") with 4 subjects per center, where each patient has a 50% chance of receiving either the drug or a placebo. A nested loop can be used to solve this problem. In the outer loop, the *index-variable*, CENTER, is assigned to the values with the name of the three cancer centers. For each iteration of the outer loop, there is an inner loop that is used to assign each patient to a group.

Program 5.7:

```
data ex5_7;
    length center $4;
    do center = "COH", "UCLA", "USC";      /* OUTER LOOP */
        do id = 1 to 4;                     /* INNER LOOP */
            if ranuni(2) > 0.5 then group = 'D';
            else group = 'P';
            output;
        end;
    end;
run;

title 'Using nested iterative loops';
proc print data = ex5_7;
run;
```

Output from Program 5.7:

```
          Using nested iterative loops
          Obs      center    id    group
            1      COH        1      P
            2      COH        2      D
            3      COH        3      D
            4      COH        4      D
            5      UCLA       1      D
            6      UCLA       2      D
            7      UCLA       3      P
            8      UCLA       4      P
            9      USC        1      P
           10      USC        2      P
           11      USC        3      D
           12      USC        4      P
```

5.1.4 Combining Implicit and Explicit Loops

In Program 5.7, all the observations were created from one DATA step since the DATA step didn't read in any input data. The CENTER variable gets its value from an outer loop. Sometimes it is necessary to use an explicit loop to create multiple observations for each observation that is read from an input data set. For example, suppose the values for CENTER are stored in a SAS data set. For each CENTER, you need to assign four patients where each patient has a 50% chance of receiving either the drug or a placebo. In this situation, you need to read in the value for the CENTER variable via the implicit loop. Then for each CENTER that is being read into the PDV, you need to utilize an explicit loop to create the ID and GROUP variables. Program 5.8 shows how an implicit and an explicit loop used together deliver what's needed.

CANCER_CENTER:

	CENTER
1	COH
2	UCLA
3	USC

Program 5.8:

```
data ex5_8;
    set cancer_center;      /* SET FOR AN IMPLICIT LOOP */
    do id = 1 to 4;         /* DO FOR AN EXPLICIT LOOP */
        if ranuni(2) > 0.5 then group = 'D';
        else group = 'P';
        output;
    end;
run;
```

In Program 5.8, the implicit loop of the DATA step is used to read in observations from the input data set, CANCER_CENTER. For each center being read, there is an explicit loop to assign patients with either 'D' or 'P.'

5.2 Utilizing Loops to Create Samples

In some situations, you may want to draw samples from a data set. There are two kinds of sampling schemes: systematic and random samples. The examples for creating samples are adapted from *SAS® Certification Prep Guide: Advanced Programming for SAS 9* (2007).

5.2.1 Direct-Access Mode

By default, SAS sequentially reads one observation per iteration of the DATA step. This process will stop when an end-of-file marker is reached. Instead of reading data sequentially, SAS can also access an observation directly via direct-access mode.

There are three important components for using the direct-access mode. First, you need to tell SAS which observation you would like to select. This step is performed by using the POINT= in the SET statement, which has the following form:

```
SET input-data-set POINT=variable;
```

The *variable* specified by the POINT= option is a temporary variable, and it is not output to the output data set. This variable is set to 0 in the PDV at the very beginning of the DATA step. Then it must be assigned to an observation number before the SET statement.

When using direct-access mode, SAS will not be able to detect the end-of-file marker. Without telling SAS explicitly when to stop processing, it will cause infinite looping. Therefore, in order to utilize the direct-access mode, you need to use the **STOP** statement at the end of the DATA step:

```
STOP;
```

Recall that SAS writes observations from the PDV to the output data set at the end of the DATA step if there is no explicit OUTPUT statement in the DATA step. However, if you use the STOP statement, the DATA step processing will stop *before* the end of the DATA step. Thus, the last step to using direct-access mode is to write an explicit OUTPUT statement before the STOP statement. Program 5.9 creates a data set that contains only the fifth observation of the SBP (*Systolic Blood Pressure*) data set.

SBP:

	ID	SBP
1	01	145
2	02	119
3	03	126
4	04	106
5	05	151
6	06	112
7	07	127
8	08	119
9	09	113

Program 5.9:

```
data ex5_9;
    obs_n = 5;
```

```
    set sbp point = obs_n;
    output;
    stop;
run;

title 'Select the fifth observation from SBP data set';
proc print data = ex5_9;
run;
```

Output from Program 5.9:
```
        Select the fifth observation from SBP data set
              Obs         id        sbp
               1          05        151
```

5.2.2 Creating a Systematic Sample

A systematic sample is created by selecting every *k*th observation from an original data set. In other words, the systematic sample cannot be created sequentially; hence, a direct-access mode must be used. You can create a systematic sample by using an iterative DO loop, which requires providing *start*, *stop*, and *increment* values. Normally, *start* is set to 1, *stop* is set to the total number of observations from the input data set, and *increment* is set to *k* that indicates every *k*th observation that you want to select.

You can determine the total number of observations by running the CONTENTS procedure. Instead of running a separate procedure, you can also provide the total number of observations by using the NOBS= option in the SET statement. Here's the general form for the NOBS= option:

```
SET input-data-set NOBS=variable;
```

The *variable* specified by the NOBS= option is a temporary variable that contains the number of observations in the *input-data-set*. It will not be written to the final data set. The *variable* is created automatically based on the descriptor portion of the *input-data-set* during the compilation phase. It will retain its value throughout the execution phase.

Program 5.10 creates a systematic sample that contains every third observation from the data set SBP. Figures 5.3 and 5.4 illustrate the execution process in detail. Notice that the automatic variable _N_ is 1 throughout the execution phase because SAS didn't read the input data sequentially. For each iteration of the DO loop, SAS uses the direct-access mode to read observations based on the observation number supplied by the CHOOSE variable. (*Program 5.10* is illustrated in *program5.10.pdf*.)

Program 5.10:

```
data ex5_10;
    do choose = 1 to total by 3;
        set sbp point = choose nobs = total;
```

```
        output;
    end;
    stop;
run;


title 'Select every third observation from SBP data set';
proc print data = ex5_10;
run;
```

Output from Program 5.10:

```
        Select every third observation from SBP data set
                Obs        id      sbp
                 1         01      145
                 2         04      106
                 3         07      127
```

5.2.3 Creating a Random Sample with Replacement

A random sample is created from an original data set on a random basis. A random sample with replacement means that an observation is returned to the original data set after it has been chosen. Hence, any observations can

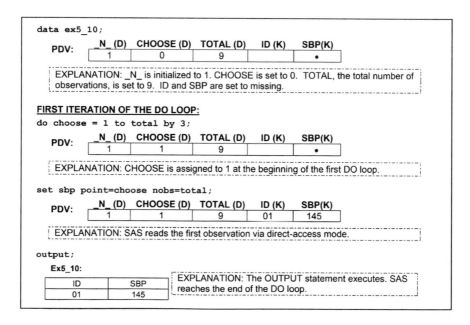

FIGURE 5.3
The first iteration of the DO loop in Program 5.10.

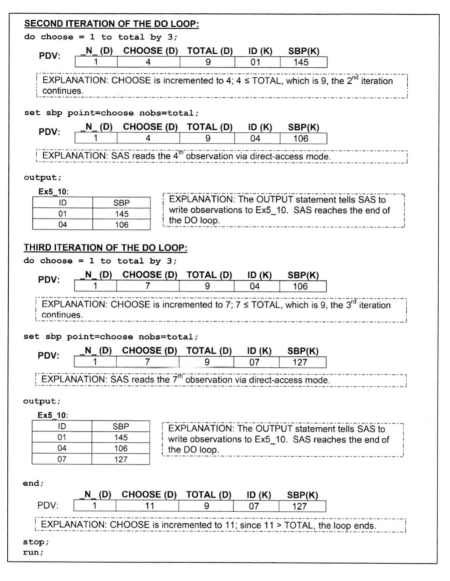

FIGURE 5.4
The last two iterations of the DO loop in Program 5.10.

be chosen more than once. Program 5.11 creates a sample of three observations with replacement from the data set SBP.

Program 5.11:

```
data ex5_11(drop = i);
    do i = 1 to 3;
```

```
        choose = ceil(ranuni(5)*total);
        set sbp point = choose nobs = total;
        output;
    end;
    stop;
run;


title 'Select a random sample with replacement';
proc print data = ex5_11;
run;
```

Output from Program 5.11:
```
        Select a random sample with replacement
                    Obs      id       sbp
                     1       09       113
                     2       08       119
                     3       09       113
```

An important step in creating a random sample with replacement is generating a random number that indicates which observation needs to be selected. In Program 5.11, to select observations randomly, an integer between 1 and the total number of observations needs to be generated within each iteration of the loop. Because the RANUNI function generates a number between 0 and 1, when multiplying this generated number, ranuni(5), with the total number of observations (TOTAL), the resulting number will be between 0 and the number of observations. Because you need an integer value, you can use the CEIL function, which returns the smallest integer that is greater than or equal to its argument.

5.2.4 Creating a Random Sample without Replacement

A random sample *without* replacement means that once an observation is randomly selected, it cannot be replaced back into the original data set. Thus, any observation cannot be chosen more than once. The algorithm to generate a random sample without replacement is more complicated than the one with replacement. The example in Program 5.12 is adapted from "SAS Programming II: Manipulating Data in the Data Step."

The algorithm for Program 5.12 is summarized in Figure 5.5. Rannum is the random number that is generated for each observation. Size is the sample size, which is decremented by 1 for each observation that is being selected for the final sample. Left is used to keep track of the total number of observations and decremented by 1 once an observation has been processed. Pct is calculated by Size divided by Left. An observation with Rannum that is less than Pct will be chosen for the final sample. The loop

	ID	SBP		RANDNUM	SIZE	LEFT	PCT = SIZE/LEFT	RANNUM < PCT?	CHOOSE
1	01	145	←	0.22	3	9	0.33	YES	1
2	02	119	←	0.64	2	8	0.25	NO	NO
3	03	126	←	0.79	2	7	0.29	NO	NO
4	04	106	←	0.11	2	6	0.33	YES	4
5	05	151	←	0.06	1	5	0.2	YES	5
6	06	112					Loop Stop!		
7	07	127							
8	08	119							
9	09	113							

FIGURE 5.5
The algorithm for Program 5.12.

stops when Size reaches 0. A DO WHILE loop will be suitable for this problem.

Program 5.12:

```
data ex5_12 (drop = pct size left randnum);
    size = 3;
    left = total;
    do while (size > 0);
        choose + 1;
        randnum = ranuni(12);
        pct = size/left;
        if randnum < pct then do;
            set sbp point = choose nobs = total;
            output;
            size = size - 1;
        end;
        left = left - 1;
    end;
    stop;
run;

title 'Select a random sample without replacement';
proc print data = ex5_12;
run;
```

Output from Program 5.12:

```
    Select a random sample without replacement
            Obs     id      sbp
             1      01      145
             2      04      106
             3      05      151
```

5.3 Using Looping to Read a List of External Files

5.3.1 Using an Iterative DO Loop to Read an External File

To read an external file, you can use the INFILE statement. Program 5.13 reads the external file, text1.txt, which is located in "C:\".

TEXT1.TXT:

```
01 145
02 119
```

Program 5.13:

```
data ex5_13;
    infile "C:\text1.txt";
    input id $ sbp;
run;
```

Because there are two observations in the external file, SAS will use two iterations in the implicit loop of the DATA step to read the data. When SAS reaches the end-of-file marker, it stops reading.

    Alternatively, you can use an explicit loop to read an external file. In order to construct an explicit loop, you need to either specify the number of iterations for the iterative DO loop or you need to specify a condition for the DO WHILE or DO UNTIL loops. One way to specify a condition is by telling SAS to read the observations until it reads the last record. In order for SAS to detect it has read the last record, you can create a temporary variable by using the END= option in the INFILE statement, which has the following form:

INFILE file-specification **END**=variable

The *variable* specified by the END= option is set to 1 when SAS reads the last record of the external file; otherwise it sets to 0. The *variable* created from the END= option is not sent to the output data set. Furthermore, you cannot use the END= option with either the DATALINES or the DATALINES4 statements.

Program 5.14:

```
data ex5_14;
    infile "C:\text1.txt" end = last;
    do until(last = 1);
        input id $ sbp;
        output;
    end;
run;
```

In Program 5.14, the DATA step execution iterates only once. Within this single iteration, the DO UNTIL loop iterates twice to read the two observations in TEXT1.TXT. During the first DATA step iteration, the automatic variable _N_ is set to 1. When SAS begins the second iteration of the DATA step, _N_ is incremented to 2 and SAS learns that the end-of-file marker has been reached. At this point the DATA step ceases its processing. Checking LAST does not occur, however, until the end of the DO UNTIL loop so that the second record still gets sent to the output data set, EX5_14. To reiterate, because the relationship between the two looping structures *is* confusing: *the end-of-file marker is detected by the DATA step, not by the DO UNTIL loop.*

5.3.2 Using an Iterative DO-Loop to Read Multiple External Files

Before processing reading multiple external files, one more option of the INFILE statement needs to be reviewed. In the INFILE statement, you generally specify the name and the location of the external file immediately after the keyword INFILE. Alternatively, you can use the FILEVAR= option in the INFILE statement to read an external file that is specified by the FILEVAR= option. The FILEVAR= option has the following form:

INFILE file-specification **FILEVAR**=variable

Like any automatic variable, the *variable* specified in the FILEVAR= option is not sent to the final data set. This *variable* contains the name of the external file and must be created before the INFILE statement. When you use the FILEVAR= option, the *file-specification* is just a placeholder, not an actual filename. For example, Program 5.15 reads an external file by using the FILEVAR= option.

Program 5.15:

```
data ex5_15;
    filename = "C:\text1.txt";
    infile dummy filevar = filename;
    input id $ sbp;
run;
```

Log from Program 5.15:
```
95    data ex5_15;
96        filename = "C:\text1.txt";
97        infile dummy filevar = filename;
98        input id $ sbp;
99    run;
NOTE: The infile DUMMY is:
        Filename = C:\text1.txt,
        RECFM = V,LRECL = 256,File Size (bytes) = 16,
        Last Modified = 04Apr2012:08:20:46,
        Create Time = 04Apr2012:08:20:34
```

```
NOTE: 2 records were read from the infile DUMMY.
      The minimum record length was 6.
      The maximum record length was 6.
NOTE: The data set WORK.EX5_15 has 2 observations and 2
      variables.
NOTE: DATA statement used (Total process time):
      real time         0.03 seconds
      cpu time          0.00 seconds
```

Notice that in the SAS log, SAS uses the placeholder, *dummy*, to report the name of the external file being read.

There are situations when you may want to read a group of external files into SAS and concatenate them into one data set. Each of the external files has identical formats. Instead of reading them individually by using separate DATA steps, you can read them all by using the FILEVAR= option in the INFILE statement in a single DATA step.

The FILEVAR= option will cause the INFILE statement to close the current input file and to open a new one, which is obtained from the FILEVAR= option. The solution for this problem is based on an example in *SAS® Certification Prep Guide: Advanced Programming for SAS 9* (2007).

For example, suppose that you would like to read TEXT1.TXT, TEXT2.TXT, and TEXT3.TXT. Both TEXT2.TXT and TEXT3.TXT have identical formats as TEXT1.TXT.

TEXT2.TXT:

```
03 126
04 106
```

TEXT3.TXT:

```
05 140
06 118
```

To read multiple files in a single DATA step, additional issues need to be addressed before coding the DATA step. In Program 5.15, three statements are used to read a single external file: the FILENAME, INFILE, and INPUT statements. In Program 5.16, these three statements need to be placed inside a loop. Notice that by naming the external files TEXT1.TXT, TEXT2.TXT, and TEXT3.TXT, you can create an iterative DO loop and iterate between 1 and 3. Within the DO loop, create the variable, NEXT, that will contain the name of the external file by concatenating "C:\text", loop index, and ".txt" via the concatenation (||) operator. If you use the index variable that contains numeric values between 1 and 3, you can also use the PUT function to convert the numeric values into character strings. For example,

```
next = "C:\text" || put (i, 1.) || ".txt";
```

If the assignment statement for NEXT is confusing, don't worry. The || operator
and the PUT function are covered more in depth in Chapter 9.

An explicit OUTPUT statement is also necessary to write the current contents from the PDV to the output data set within the loop. Because you are
using the FILEVAR= option to control closing the current input file and open-
ing a new file, SAS will not be able to detect the end-of-file marker. Thus, you
also need to place a STOP statement outside the loop. In order to see which
observation is read from which file, a FILENAME variable is also created in
Program 5.16.

Program 5.16:

```
data ex5_16 (drop = i);
    do i = 1 to 3;
        next = "C:\text" || put (i, 1.) || ".txt";
        infile temp filevar = next;
        /* THE AUTOMATIC VARIABLE NEXT IS ASSIGNED TO DATA SET
            VARIABLE FILENAME */
        fileName = next;
        input id $ sbp;
        output;
    end;
    stop;
run;


title 'Reading multiple external files - first attempt';
proc print data = ex5_16;
run;
```

Output from Program 5.16:

```
        Reading multiple external files - first attempt
            Obs        fileName        id    sbp
            1        C:\text1.txt      01    145
            2        C:\text2.txt      03    126
            3        C:\text3.txt      05    140
```

Program 5.16 didn't create the output that we expected; only the first
observation from each external file was sent to the designated output data set. The INPUT statement within the loop only read one line
of the external file. When a single iteration of the DO loop has been
completed, the following iteration starts to read from a new file that is
specified by the NEXT variable. In order to read all the observations
of each file, you can utilize the END= option and include a DO UNTIL
loop within the iterated DO loop. Program 5.17 is a modified program
and reads the data correctly. A detailed explanation of Program 5.17 is
illustrated in Figures 5.6 through 5.8. (*Program 5.17* is also illustrated in
program5.17.pdf.)

```
data ex5_17(drop=i);
```

_N_ (D)	I (D)	NEXT (D)	LAST(D)	ID(K)	SBP(K)
1	•		0		•

EXPLANATION: _N_ is initialized to 1, LAST is initialized to 0. Others are set to missing.

FIRST ITERATION OF THE DO LOOP (OUTER LOOP):

```
do i = 1 to 3;
```

_N_ (D)	I (D)	NEXT (D)	LAST(D)	ID(K)	SBP(K)
1	1		0		•

EXPLANATION: I is set to 1 at the beginning of the first DO loop.

```
next = "C:\text" || put(i, 1.) || ".txt";
```

_N_ (D)	I (D)	NEXT (D)	LAST(D)	ID(K)	SBP(K)
1	1	C:\text1.txt	0		•

EXPLANATION: NEXT is assigned with 'C:\text1.txt'.

FIRST ITERATION OF THE DO UNTIL LOOP (INNER LOOP):

```
do until (last);
```
EXPLANATION: The DO UNTIL loop evaluates the condition at the end of the loop.

```
infile temp filevar=next end=last;
```

	1	2	3	4	5	6	7	...
Input buffer:	0	1			1	4	5	...

EXPLANATION: INFILE statement reads the first data line from 'text1.txt' into the input buffer.

```
input id $ sbp;
```

_N_ (D)	I (D)	NEXT (D)	LAST(D)	ID(K)	SBP(K)
1	1	C:\text1.txt	0	01	145

EXPLANATION: INPUT statement reads data from the input buffer to the PDV.

```
output; end;
```

Ex5_17:

ID	SBP
01	145

EXPLANATION: The OUTPUT statement executes. SAS reaches the end of the DO UNTIL loop. Since LAST ≠1, the inner loop continues.

SECOND ITERATION OF THE DO UNTIL LOOP (INNER LOOP):

```
do until (last); infile temp filevar=next end=last;
```

	1	2	3	4	5	6	7	...
Input buffer:	0	2			1	1	9	...

EXPLANATION: INFILE statement reads the second data line from 'text1.txt' into the input buffer. LAST is set to 1 in the PDV.

```
input id $ sbp;
```

_N_ (D)	I (D)	NEXT (D)	LAST(D)	ID(K)	SBP(K)
1	1	C:\text1.txt	1	02	119

EXPLANATION: INPUT reads data values from the input buffer and writes them to the PDV.

```
output; end;
```

Ex5_17:

ID	SBP
01	145
02	119

EXPLANATION: The OUTPUT statement executes. SAS reaches the end of the DO UNTIL loop. Since LAST = 1, the inner loop ends.

```
end;
```
EXPLANATION: SAS reaches the end of outer loop.

FIGURE 5.6
The first iteration of the DO loop in Program 5.17.

SECOND ITERATION OF THE DO LOOP (OUTER LOOP):

```
do i = 1 to 3;
```

_N_ (D)	I (D)	NEXT (D)	LAST(D)	ID(K)	SBP(K)
1	2	C:\text1.txt	1	02	119

EXPLANATION: I is assigned to 2; since I ≤ 3, the second iteration of the outer loop continues.

```
next = "C:\text" || put(i, 1.) || ".txt";
```

_N_ (D)	I (D)	NEXT (D)	LAST(D)	ID(K)	SBP(K)
1	2	C:\text2.txt	1	02	119

EXPLANATION: The NEXT variable is assigned with the value 'C:\text2.txt'.

FIRST ITERATION OF THE DO UNTIL LOOP (INNER LOOP):

```
do until (last); infile temp filevar=next end=last;
```

Input buffer:

1	2	3	4	5	6	7	...
0	3		1	2	6		...

EXPLANATION: INFILE reads 1st data line from 'text2.txt' into the input buffer. This is not the last record of 'text2.txt'; LAST is set to 0 in the PDV.

```
input id $ sbp;
```

_N_ (D)	I (D)	NEXT (D)	LAST(D)	ID(K)	SBP(K)
1	2	C:\text2.txt	0	03	126

EXPLANATION: The INPUT statement reads data values from the input buffer to the PDV.

```
output; end;
```

Ex5_17:

ID	SBP
01	145
02	119
03	126

EXPLANATION: The OUTPUT statement executes. SAS reaches the end of the DO UNTIL loop. Since LAST ≠1, the inner loop continues.

SECOND ITERATION OF THE DO UNTIL LOOP (INNER LOOP):

```
do until (last); infile temp filevar=next end=last;
```

Input buffer:

1	2	3	4	5	6	7	...
0	4		1	0	6		...

EXPLANATION: INFILE reads the 2nd data line into the input buffer. This is the last record of 'text2.txt', LAST is set to 1 in the PDV.

```
input id $ sbp;
```

_N_ (D)	I (D)	NEXT (D)	LAST(D)	ID(K)	SBP(K)
1	2	C:\text2.txt	1	04	106

EXPLANATION: INPUT statement reads data values from the input buffer and writes them to the PDV.

```
output; end;
```

Ex5_17:

ID	SBP
01	145
02	119
03	126
04	106

EXPLANATION: The OUTPUT statement executes. SAS reaches the end of the DO UNTIL loop. Since LAST = 1, the inner loop ends.

```
end;
```

EXPLANATION: SAS reaches the end of outer loop

FIGURE 5.7
The second iteration of the DO loop in Program 5.17.

TRIRD ITERATION OF THE DO LOOP (OUTER LOOP):

```
do i = 1 to 3;
```

N (D)	I (D)	NEXT (D)	LAST(D)	ID(K)	SBP(K)
1	3	C:\text2.txt	1	04	106

> EXPLANATION: I is assigned to 3; since I ≤ 3, the second iteration of the outer loop continues.

```
next = "C:\text" || put(i, 1.) || ".txt";
```

N (D)	I (D)	NEXT (D)	LAST(D)	ID(K)	SBP(K)
1	3	C:\text3.txt	1	04	106

> EXPLANATION: NEXT is assigned with 'C:\text3.txt'.

FIRST ITERATION OF THE DO UNTIL LOOP (INNER LOOP):

```
do until (last); infile temp filevar=next end=last;
```

           1 2 3 4 5 6 7 ...

Input buffer: | 0 | 5 | | 1 | 4 | 0 | | ...

> EXPLANATION INFILE reads the 1st data line from 'text3.txt' into the input buffer. This is not the last record of 'text3.txt', LAST is set to 0.

```
input id $ sbp;
```

N (D)	I (D)	NEXT (D)	LAST(D)	ID(K)	SBP(K)
1	3	C:\text3.txt	0	05	140

> EXPLANATION: INPUT statement reads data values to the PDV

```
output; end;
```

Ex5_17:

ID	SBP
01	145
02	119
03	126
04	106
05	140

> EXPLANATION: The OUTPUT statement executes. SAS reaches the end of the DO UNTIL loop. Since LAST ≠1, the inner loop continues.

SECOND ITERATION OF THE DO UNTIL LOOP (INNER LOOP):

```
do until (last);  infile temp filevar=next end=last;
```

           1 2 3 4 5 6 7 ...

Input buffer: | 0 | 6 | | 1 | 1 | 8 | | ...

> EXPLANATION INFILE reads the 2nd data line from 'text3.txt' into the input buffer. Since this is the last record of 'text3.txt', LAST is set to 1.

```
input id $ sbp;
```

N (D)	I (D)	NEXT (D)	LAST(D)	ID(K)	SBP(K)
1	3	C:\text3.txt	1	06	118

> EXPLANATION: INPUT statement reads data values to the PDV

```
output; end;
```

Ex5_17:

ID	SBP
01	145
02	119
03	126
04	106
05	140
06	118

> EXPLANATION: The OUTPUT statement executes. SAS reaches the end of the DO UNTIL loop. Since LAST = 1, the inner loop ends.

```
end;
```
> EXPLANATION: SAS reaches the end of outer loop.

> EXPLANATION: At the beginning of the fourth iteration of the outer loop, I is incremented to 4. Since I > 3, the outer loop ends.

```
stop; run;
```
> EXPLANATION: The STOP statement stops the DATA step processing.

FIGURE 5.8
The third iteration of the DO loop in Program 5.17.

Program 5.17:

```
data ex5_17(drop = i);
    do i = 1 to 3;
        next = "C:\text" || put(i, 1.)|| ".txt";
        do until (last);
            infile dummy filevar = next end = last;
            fileName = next;
            input id $ sbp;
            output;
        end;
    end;
    stop;
run;

title 'Reading multiple external files - second attempt';
proc print data = ex5_17;
run;
```

Output from Program 5.17:

```
        Reading multiple external files - second attempt
        Obs         fileName         id       sbp
         1        C:\text1.txt        01       145
         2        C:\text1.txt        02       119
         3        C:\text2.txt        03       126
         4        C:\text2.txt        04       106
         5        C:\text3.txt        05       140
         6        C:\text3.txt        06       118
```

Exercises

Exercise 5.1. Suppose that you would like to invest $1,000 each year at a bank. The investment earns 5% annual interest, compounded monthly (that means the interest for each month will be 0.05/12). Write a program using an explicit loop (or loops) to calculate your balance for each month if you are investing for 2 years (that means you deposit $1,000 at the beginning of each year).

Exercise 5.2. A quadratic equation ($ax^2 + bx + c = 0$) is a polynomial equation of the second degree. The solution for the quadratic equation is as follows:

$$x = \frac{-b \pm \sqrt{b^2 - 4ac}}{2a}$$

In the above formula, the expression within the square root sign ($b^2 - 4ac$) is the discriminant of the quadratic equation. If the discriminant is positive ($b^2 - 4ac > 0$), then there are two distinct real solutions. If the discriminant is zero ($b^2 - 4ac = 0$), then there is exactly one real root solution, which is $-b/2a$. If the discriminant is negative ($b^2 - 4ac < 0$), then there is no real solution. The following program illustrates how to calculate the solutions for the quadratic equation.

```
data example;
    a = 1;
    b = 3;
    c = 2;
    x1 = ((-1*b)-sqrt(b**2-4*a*c))/(2*a);
    x2 = ((-1*b)+sqrt(b**2-4*a*c))/(2*a);
run;
```

For this problem, write a program by using explicit loops to solve the quadratic equation based on the combination of the following values:

- A = -2, -1, 0, 1, 2
- B = -6, -5, -4
- C = 1, 2, 3

If A equals 0, the equation becomes a linear equation ($bx + c = 0$). The solution for the linear equation is $x = -c/b$.

The resulting data set needs to contain the following variables in addition to the A, B, and C variables:

- The variables X1 and X2, which are the solutions for the equation. If there is only one solution for either quadratic or linear equation, use X1 for the solution and leave X2 to missing.
- The variable NOTE will have either "Quadratic" (for solving the quadratic equation), "Linear" (for solving the linear equation), or "No Solution" (when discriminant is negative) values.

Exercise 5.3. To generate a random number that follows standard normal distribution (mean = 0, standard deviation = 1), you can use the RANNOR function. To generate a random number that follows a normal distribution with mean equaling *mu* and standard deviation equaling *sd*, you can use this formula: mu + sd*rannor(seed).

For this exercise, you need to simulate a data set that contains 1,000 subjects and contains three variables: ID, GROUP, and WEIGHT.

- ID: Contains values from 1 to 1,000.
- GROUP: Each subject has a 50% chance of being assigned to either the 'A' or 'B' groups.
- WEIGHT: For subjects assigned to group 'A,' their weight will be generated by the following normal distribution with mean equal to 200 and standard deviation equal to 15. For subjects assigned to group 'B,' their mean weight will be 250 and standard deviation will equal 50.

After you simulate the data, use the appropriate procedure to verify that the GROUP and WEIGHT variables were correctly generated.

6

Array Processing

6.1 Introduction to Array Processing

If you are familiar with other programming languages, you have probably worked with arrays. Arrays allow you to modify data more efficiently because you can use them to perform the same tasks for a group of related variables inside a looping structure.

Working with arrays, however, is more complicated in SAS®, partly because the syntax has been enriched to allow for nuanced data processing. Even when the syntax is mastered, one still needs to understand how the array interacts with what is going on in the PDV. Knowing when it is best to replace existing code with an array presents an additional challenge. In this chapter, a wide range of applications that use arrays in looping structures will be covered. When you finish reading this chapter, you should know how arrays work and when best to use them in your own SAS programs. The material in this chapter is based upon a paper that I presented at SAS Global Forum (Li, 2011).

6.1.1 Situations for Utilizing Array Processing

The following example illustrates a typical situation for utilizing array processing. The data set SBP contains six measurements of systolic blood pressure (SBP) measurements for four patients. The missing values are coded as 999. Suppose that you would like to recode 999 to the SAS value for missing (.). You may want to write the code like the one in Program 6.1.

SBP:

	SBP1	SBP2	SBP3	SBP4	SBP5	SBP6
1	141	142	137	117	116	124
2	999	141	138	119	119	122
3	142	999	139	119	120	999
4	136	140	142	118	121	123

One dimensional array SBPARY:

SBP1	SBP2	SBP3	SBP4	SBP5	SBP6
[1]	[2]	[3]	[4]	[5]	[6]

Two dimensional array SBPARY2:

	[1]	[2]	[3]
[1]	SBP1	SBP2	SBP3
[2]	SBP4	SBP5	SBP6

FIGURE 6.1
The conceptual view of one- and two-dimensional arrays.

Program 6.1:

```
data ex6_1;
    set sbp;
    if sbp1 = 999 then sbp1 = .;
    if sbp2 = 999 then sbp2 = .;
    if sbp3 = 999 then sbp3 = .;
    if sbp4 = 999 then sbp4 = .;
    if sbp5 = 999 then sbp5 = .;
    if sbp6 = 999 then sbp6 = .;
run;
```

Each of the IF statements in Program 6.1 converts the number 999 to a missing SAS value. These IF statements are almost identical; only the names of the variables are different. If these six variables can be grouped into a one-dimensional array (see Figure 6.1), then you can recode the variables in a DO loop. In this situation, grouping variables into an array will make code writing more efficient.

In Figure 6.1, variables SBP1 to SBP6 are grouped into a one-dimensional array, SBPARY. Once these variables are grouped into an array, you can reference the data in these variables by using the SBPARY[*n*] format, where *n* is the element number. For example, to reference the data in the SBP3 variable, you can use SBPARY[3].

Variables in the DATA step can also be grouped into multi-dimensional arrays. For example, in the bottom portion of Figure 6.1, SBP1 to SBP6 are grouped into a two-dimensional array, SBPARY2. To reference the data in the variables in SBPARY2, you need to use the SBPARY2[*r, c*] format, where *r* is the row number and *c* is the column number. For example, to reference the data in the SBP3 variable, you can use SBPARY2[1,3]. To reference the data in SBP5, use SBPARY2[2,2].

6.1.2 Defining and Referencing One-Dimensional Arrays

A SAS array is a temporary grouping of SAS variables under a single name. Arrays exist only for the duration of the DATA step. The ARRAY statement is used to either group previously defined variables or to create a group of variables, which has the following form:

ARRAY array-name [subscript] <$><length><array-elements>
        < (initial-value-list) >;

The array-name in the ARRAY statement does not become part of the output data. The *array-name* must be a legitimate SAS name and cannot be the name of a SAS variable in the same DATA step. If you happen to use a function name as *array-name*, this name would be treated as the name of an array in parenthetical references, not the name of the function. SAS will also send a warning message to your SAS log. Furthermore, *array-name* cannot be used in the LABEL, FORMAT, DROP, KEEP, or LENGTH statements.

The [subscript] component in the ARRAY statement describes the number or arrangement of array elements and can be specified in different forms. You can enclose *subscript* with braces {}, brackets [], or parentheses (). The simple form for [subscript] is to specify the dimensional size of the array. For example, you can group SBP1 to SBP6 into an array like below:

```
array sbpary[6] sbp1 sbp2 sbp3 sbp4 sbp5 sbp6;
```

You can also provide a range of numbers as [subscript] by providing the lower and upper bounds of the array and separate them by a colon (:). For example,

```
array sbpary[1990:1993] sbp1990 sbp1991 sbp1992 sbp1993;
```

An asterisk (*) can also be used as [subscript]. Using an asterisk will let SAS determine the subscript by counting the variables in the array. When you specify the asterisk, you must include *array-elements*. For example,

```
array sbpary[*] sbp1 sbp2 sbp3 sbp4 sbp5 sbp6;
```

The optional $ (dollar sign) indicates that the elements in the array are character elements. You don't need to specify the $ if the array elements have been previously defined as character elements. If the lengths of array elements have not been previously specified, you can use the *length* option in the ARRAY statement.

The optional *array-elements* are the variables to be included in the array, which must either be all numeric or character variables. If variable names that are included in the array end with consecutive numeric values, you can use a shorthand notation by connecting the first and last variables with a hyphen. For example, SBP1-SBP6 is equivalent to listing SBP1 to SBP6 individually.

If *array-elements* are not specified, the *array-elements* would be implied to the variables with the names that contain array names and the numbers 1, 2, … n. For example,

```
array sbp[6];
```

is equivalent to the following statement:

```
array sbp[6] sbp1-sbp6;
```

When *array-elements* (SBP1-SBP6) are not listed, SBP1 to SBP6 can already exist in the DATA step. However, if SBP1 to SBP6 do not already exist, they will be created in the DATA step.

Another way to list *array-elements* is to use the keywords _NUMERIC_, _CHARACTER_, and _ALL_, which are used to specify all numeric, all character, or all the same type of variables. If the keyword _ALL_ is used, all the previously defined variables must have the same type. For example,

```
array num[*] _numeric_;
array char[*] _character_;
array allvar[*] _all_;
```

You can use the keyword _TEMPORARY_ as *array-elements* to create temporary arrays. Using temporary arrays is useful when you want to create an array only for computing purposes. When referring to a temporary data element, you refer to it by the *array-name* and its dimension. Because the temporary array contains only constants as elements, they cannot be sent as variables to the output data set. Also, the values in temporary arrays are automatically retained without being reset to missing at the beginning of each iteration of the DATA step execution. You cannot use asterisks (*) with temporary arrays.

You can also assign initial values to the array elements in the (*initial-value-list*) when creating a group of variables by using the ARRAY statement. The initial values can also be assigned to temporary data elements. The values can be either numbers or character strings separated by either a comma or a blank space. When any or all elements in an array are assigned with initial values, all elements in the array would act as they are named in a RETAIN statement. For example, the following ARRAY statement creates N1, N2, and N3 DATA step variables by using the ARRAY statement and initializes them with the values 1, 2, and 3, respectively:

```
array num[3] n1 n2 n3 (1 2 3);
```

The following ARRAY statement creates CHR1, CHR2, and CHR3 variables. The dollar sign ($) is necessary because CHR1, CHR2, and CHR3 are not previously defined in the DATA step.

```
array chr[3] $ ('A', 'B', 'C');
```

Next, the ARRAY statement creates a temporary array, NUM, and the number of elements in the array is 3. Each element in the array is initialized to 1, 2, and 3:

```
array num[3] _temporary_ (1 2 3);
```

Again, notice that NUM1, NUM2, and NUM3 are not created as variables, because a temporary array is being defined.

After an array is defined, you can use the ARRAY reference statement to describe the elements in an array to be processed in the DATA step. The ARRAY reference statement has the following form:

```
array-name[subscript]
```

The *array-name* is the name of an array that was previously defined within the sample DATA step. The [*subscript*] component is used to specify the subscript of the array. You can use the name of a variable, an asterisk (*), or a SAS expression as the *subscript*.

When a variable is used as the *subscript*, the array referencing it is used with an iterative DO loop. Within each iteration of execution of the DO loop, the current value of this variable is used as the *subscript* of the array element being processed in the DATA step.

When an asterisk (*) is used as the *subscript*, SAS will treat the elements in the array as a variable list. You cannot reference the temporary array elements with an asterisk.

When a SAS expression is used as *subscript* in a SAS statement, the expression needs to be evaluated to a value when the statement executes. The evaluated value needs to be within the lower and upper bounds of the array. You can also use an integer value as *subscript*. Program 6.2 is a modified version of Program 6.1 by using array processing.

Program 6.2:

```
data ex6_2(drop = i);
    set sbp;
    array sbpary[6] sbp1-sbp6;
    do i = 1 to 6;
        if sbpary[i] = 999 then sbpary[i] = .;
    end;
run;
```

6.1.3 Compilation and Execution Phases for Array Processing

During the compilation phase, the PDV is created (_ERROR_ is omitted for simplicity). The array name SBPARY and array references are not included in the PDV. (See Figure 6.2.) Each variable, SBP1 to SBP6, is referenced by

_N_	SPB1(K)	SPB2(K)	SPB3(K)	SPB4(K)	SPB5(K)	SPB6(K)	I (D)

```
        ↑           ↑          ↑          ↑          ↑          ↑
   SBPARY[1]  SBPARY[2]  SBPARY[3]  SBPARY[4]  SBPARY[5]  SBPARY[6]
```

FIGURE 6.2
The array references are not included in the program data vector (PDV) during the compilation phase.

the array reference. Syntax errors in the ARRAY statement will be detected during the compilation phase.

The first two iterations of the DATA step execution are illustrated in Figures 6.3 and 6.4.

6.2 Functions and Operators Related to Array Processing

6.2.1 The DIM, HBOUND, and LBOUND Functions

Sometimes you won't know the dimension of an array. This tends to be the case when you use _NUMERIC_, _CHARACTER_, and _ALL_ as *array-elements*. A handy work-around is the DIM function, which supplies the actual dimension to the DATA step. Closely related to the DIM function are the HBOUND and LBOUND functions that return the upper and lower bounds of an array, respectively. The three functions have the following forms:

```
DIM<n>(array-name)
DIM(array-name,bound-n)
HBOUND<n>(array-name)
HBOUND(array-name,bound-n)
LBOUND<n>(array-name)
LBOUND(array-name,bound-n)
```

Notice that these functions share similar syntax. The optional *n* is used to specify the dimension of an array. If the *n* value is not specified, the DIM function will return the number of elements in the first dimension of the array, and HBOUND and LBOUND will return the upper bound and lower bound of the first dimension, respectively. The *bound-n* is a numeric constant, variable, or SAS expression used to specify the dimension for which you want to know the number of elements in a multi-dimensional array from the DIM function. If *bound-n* is used in

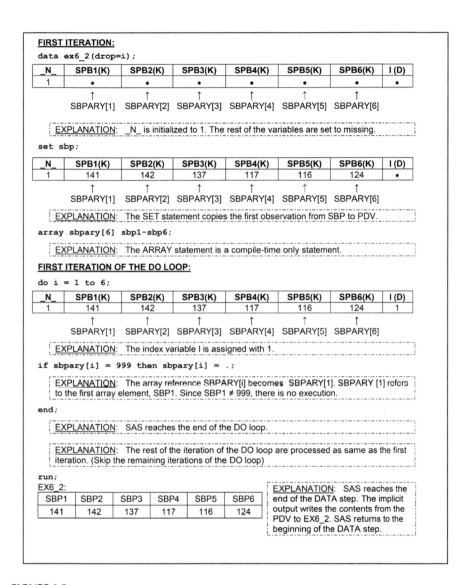

FIGURE 6.3
The first iteration of the DATA step in Program 6.2.

the HBOUND or LBOUND functions, the upper or lower bound of the array will be returned. You can only use *bound-n* when *n* is not specified. For example, in Program 6.3 the DIM function is used to return the number of elements in SBPARY. Instead of using the DIM function, you can also use the HBOUND function to return the upper bound of SBPARY, which is 6.

SECOND ITERATION:

```
data ex6_2(drop=i);
```

N	SPB1(K)	SPB2(K)	SPB3(K)	SPB4(K)	SPB5(K)	SPB6(K)	I (D)
2	141	142	137	117	116	124	•

    ↑   ↑   ↑   ↑   ↑   ↑
SBPARY[1] SBPARY[2] SBPARY[3] SBPARY[4] SBPARY[5] SBPARY[6]

EXPLANATION: _N_ is incremented to 2. SBP1 – SBP6 retained their values. I is set to missing since I is not from the input data set.

```
set sbp; array sbpary[6] sbp1-sbp6;
```

N	SPB1(K)	SPB2(K)	SPB3(K)	SPB4(K)	SPB5(K)	SPB6(K)	I (D)
2	999	141	138	119	119	122	•

SBPARY[1] SBPARY[2] SBPARY[3] SBPARY[4] SBPARY[5] SBPARY[6]

EXPLANATION: The SET statement copies the second observation from SBP to PDV.

FIRST ITERATION OF THE DO LOOP:

```
do i = 1 to 6;
```

N	SPB1(K)	SPB2(K)	SPB3(K)	SPB4(K)	SPB5(K)	SPB6(K)	I (D)
2	999	141	138	119	119	122	1

SBPARY[1] SBPARY[2] SBPARY[3] SBPARY[4] SBPARY[5] SBPARY[6]

EXPLANATION: The index variable I is assigned with 1.

```
if sbpary[i] = 999 then sbpary[i] = .;
```

EXPLANATION: The array reference SBPARY[i] becomes SBPARY[1]. SBPARY [1] refers to the first array element, SBP1. Since SBP1 = 999, SBP1 is set to *missing*

N	SPB1(K)	SPB2(K)	SPB3(K)	SPB4(K)	SPB5(K)	SPB6(K)	I (D)
2	•	141	138	119	119	122	1

SBPARY[1] SBPARY[2] SBPARY[3] SBPARY[4] SBPARY[5] SBPARY[6]

```
end;
```

EXPLANATION: SAS reaches the end of the DO loop.

EXPLANATION: The rest of the iteration of the DO loop are processed as same as the first iteration. (Skip the remaining iterations of the DO loop)

```
run;
```

EX6_2:

SBP1	SBP2	SBP3	SBP4	SBP5	SBP6
141	142	137	117	116	124
•	141	138	119	119	122

EXPLANATION: SAS reaches the end of the DATA step. The implicit output writes the contents from the PDV to EX6_2. SAS returns to the beginning of the DATA step.

FIGURE 6.4
The second iteration of the DATA step in Program 6.2.

Program 6.3:

```
data ex6_3(drop = i);
    set sbp;
    array sbpary[*] _numeric_;
    do i = 1 to dim(sbpary); /*Or: do i = 1 to hbound(sbpary)*/
```

```
        if sbpary[i] = 999 then sbpary[i] =.;
    end;
run;
```

LBOUND and HBOUND are typically used when the lower bound of an array dimension has a value other than 1 or the upper bound has a value other than *n*.

6.2.2 Using the IN and OF Operator with an Array

The data set SBP2 is similar to the data set SBP except that SBP2 contains the data with the correct numerical missing values:

SBP2:

	SBP1	SBP2	SBP3	SBP4	SBP5	SBP6
1	141	142	137	117	116	124
2	.	141	138	119	119	122
3	142	.	139	119	120	.
4	136	140	142	118	121	123

For example, suppose that you would like to create a variable, MISS, that is used to indicate whether SBP1–SBP6 contains missing values. This task can be easily accomplished by using the IN operator.

The IN operator, introduced in Chapter 1, is used to determine whether a variable's value is among the list of character or numeric values. You can use the IN operator to search for numeric or character values in an array. Program 6.4 illustrates the use of the IN operator with SBPARY to create the MISS variable.

Program 6.4:

```
data ex6_4;
    set sbp2;
    array sbpary[*] _numeric_;
    miss = . IN sbpary;
run;

title 'Using the IN operator to create variable MISS';
proc print data = ex6_4;
run;
```

Output from Program 6.4:

```
    Using the IN operator to create variable MISS
  Obs   sbp1   sbp2   sbp3   sbp4    sbp5   sbp6   miss
   1     141    142    137    117     116    124    0
```

2	.	141	138	119	119	122	1
3	142	.	139	119	120	.	1
4	136	140	142	118	121	123	0

You can pass an array on to most functions with the OF operator. Suppose that you would like to create two variables, SBP_MIN and SBP_MAX, that contain the minimum and maximum SBP values for each person. You can use the MIN and MAX function with the OF operator to accomplish this task. (See Problem 6.5.) Note that you must use an asterisk as the subscript within the function.

Program 6.5:

```
data ex6_5;
    set sbp2;
    array sbpary[*] _numeric_;
    sbp_min = min(of sbpary[*]);
    sbp_max = max(of sbpary[*]);
run;

title 'Using the OF operator to create variables SBP_MIN and
SBP_MAX';
proc print data = ex6_5;
run;
```

Output from Program 6.5:

```
Using the OF operator to create variables SBP_MIN and SBP_MAX
Obs   sbp1   sbp2  sbp3   sbp4   sbp5   sbp6  sbp_min  sbp_max
 1     141    142   137    117    116    124    116      142
 2      .     141   138    119    119    122    119      141
 3     142     .    139    119    120     .     119      142
 4     136    140   142    118    121    123    118      142
```

6.3 Some Array Applications

6.3.1 Creating a Group of Variables by Using Arrays

Suppose that the first three measurements of SBP in the SBP2 data set are recorded pre-treatment and the last three are recorded post-treatment. Furthermore, suppose that the average SBP value for the pre-treatment measurements is 140 and that the average SBP is 120 for the measurements after the treatments. You would like to create a list of variables, ABOVE1–ABOVE6, that indicate whether each measurement is above (1) or below (0) the average measurement.

The solution for this problem is in Program 6.6. The first ARRAY statement in Program 6.6 is used to group the existing variables, SBP1–SBP6. The second

ARRAY statement creates six new DATA step variables, ABOVE1 through ABOVE6. The third ARRAY statement creates temporary data elements used only for comparison purposes.

Program 6.6:

```
data ex6_6(drop = i);
    set sbp2;
    array sbp[6];
    array above[6];
    array threshold[6] _temporary_ (140 140 140 120 120 120);
    do i = 1 to 6;
        if (not missing(sbp[i]))
            then above[i] = sbp[i] > threshold[i];
    end;
run;

title 'Creating variables ABOVE1 - ABOVE6';
proc print data = ex6_6;
run;
```

Output from Program 6.6:

```
        Creating variables ABOVE1 - ABOVE6
                                       a    a    a    a    a    a
                                       b    b    b    b    b    b
         s    s    s    s    s    s    o    o    o    o    o    o
  O      b    b    b    b    b    b    v    v    v    v    v    v
  b      p    p    p    p    p    p    e    e    e    e    e    e
  s      1    2    3    4    5    6    1    2    3    4    5    6
  1    141  142  137  117  116  124    1    1    0    0    0    1
  2      .  141  138  119  119  122    .    1    0    0    0    1
  3    142    .  139  119  120    .    1    .    0    0    0    .
  4    136  140  142  118  121  123    0    0    1    0    1    1
```

6.3.2 Calculating Products of Multiple Variables

You can use the SUM function to calculate the sum of multiple variables. However, SAS does not have a built-in function to calculate the *product* of multiple variables. The easiest way to calculate the product of multiple variables is to use array processing. For example, in Program 6.7, the product of NUM1–NUM4 is calculated for each observation in the PRODUCT data set with array processing.

PRODUCT:

	NUM1	NUM2	NUM3	NUM4
1	4	.	2	3
2	.	2	3	1

Program 6.7:

```
data ex6_7 (drop = i);
    set product;
    array num[4];
    if missing(num[1]) then result = 1;
    else result = num[1];
    do i = 2 to 4;
        if not missing(num[i]) then result = result*num[i];
    end;
run;

title 'Calculating the product of NUM1-NUM4';
proc print data = ex6_7;
run;
```

Output from Program 6.7:

```
          Calculating the product of NUM1-NUM4
    Obs     num1     num2     num3     num4     result
     1       4        .        2        3        24
     2       .        2        3        1         6
```

6.3.3 Restructuring Data Sets Using One-Dimensional Arrays

Transforming a data set with one observation per subject to multiple observations per subject, or vice versa, was introduced in Sections 3.3.3 and 4.2.4. The solutions in these two sections are not efficient if you have large numbers of variables that need to be transposed. A more effective approach is to utilize array processing. The next two programs use the same data sets, WIDE and LONG, that were introduced in Chapter 3.

In Program 6.8, the variables S1–S3 are grouped into the array S, and the program utilizes the iterative DO loop with the TIME variable as the index variable (to assign S[TIME] to the SCORE variable). The output is generated within the DO loop when the SCORE variable is not missing. Program 6.8 transforms data set WIDE into a *long* format with array processing.

Program 6.8:

```
data ex6_8 (drop = s1-s3);
    set wide;
    array s[3];
    do time = 1 to 3;
        score = s[time];
        if not missing(score) then output;
    end;
run;
```

Program 6.9 does the opposite by transforming data set LONG to a *wide* format. Array processing is still used in the second transformation. Variables S1, S2, and S3 are created by using the ARRAY statement. You can retain all the elements in the array by using the RETAIN statement and the name of the array. When reading the first observation of each subject (FIRST.ID = 1), an iterative DO loop is used to initialize each element in the array (S[I]) to missing. Utilizing array processing simplifies processing shown in Program 4.5 from Chapter 4 because the cumbersome IF-THEN/ELSE statement is replaced with an assignment statement that uses TIME as the *subscript* of the S array (S[TIME] = SCORE).

Program 6.9:

```
data ex6_9 (drop = time score i);
    set long;
    by id;
    array s[3];
    retain s;
    if first.id then do;
        do i = 1 to 3;
            s[i] = .;
        end;
    end;
    s[time] = score;
    if last.id;
run;
```

6.4 Applications That Use Multi-Dimensional Arrays

The syntax for creating multi-dimensional arrays is similar to the one for creating one-dimensional arrays. The only difference is that multiple numbers or ranges of numbers appear in the array *subscript*. The numbers for each dimension are separated by a comma. The first number of the *subscript* refers to the number of rows, and the second number refers to the number of columns. If there are three dimensions, the next number will refer to the number of pages.

6.4.1 Calculating Average SBP for Pre- and Post-Treatment

Suppose that the SBP2 data set contains three pre-treatment measurements of SBP (SBP1–SBP3) and three post-treatment measurements of SBP values (SBP4–SBP6). You would like to create two variables that contain average SBP for pre- and post-treatments for each patient. One way to solve this problem is to use a two-dimensional array as in Program 6.10.

Program 6.10:

```
data ex6_10 (drop = i j);
    set sbp2;
    array sbp[2, 3];    /* Row 1 = Pre Treatment, Row 2 = Post
                           Treatment */
                        /* Columns 1-3 are SBP measurements */
    array sbpmean[2];   /* Two means per subject */
    array n[2] _temporary_;
    array sbpsum [2] _temporary_;
    do i = 1 to 2;
        sbpsum[i] = 0;
        n[i] = 0;
        do j = 1 to 3;
            sbpsum[i] + sbp[i,j];
            if not missing(sbp[i,j]) then n[i] + 1;
        end;
        sbpmean[i] = sbpsum[i]/n[i];
    end;
run;


title 'Calculating the average of SBP for pre- and post-treat-
ment';
proc print data = ex6_10;
run;
```

Output from Program 6.10:

```
Calculating the average of SBP for pre- and post-treatment
Obs   sbp1   sbp2   sbp3   sbp4   sbp5   sbp6   sbpmean1   sbpmean2
 1    141    142    137    117    116    124   140.000    119.000
 2     .     141    138    119    119    122   139.500    120.000
 3    142     .     139    119    120     .    140.500    119.500
 4    136    140    142    118    121    123   139.333    120.667
```

Program 6.10 uses a 2-by-3 two-dimensional array to group SBP1–SBP6; the first row of the array will contain the pre-treatment measurements (SBP1–SBP3), and the second row will contain the post-treatment results (SBP4–SBP6). There will be two variables that contain the mean measures for pre- and post- measurements; thus, a one-dimensional array (SBPMEAN) with two elements can be used to hold these two values for each patient. A one-dimensional array (N) is used to accumulate the non-missing measurements used to calculate the mean. A nested loop is used in the program. There are two iterations for the outer loop: one for pre-treatment measurements and one for post-treatment measurements. There are three iterations for the inner loop because there are three measurements for the pre- and post-measurements.

6.4.2 Restructuring Data Sets by Using a Multi-Dimensional Array

The DAT1 data set in this section contains two records for each person, whereas the DAT2 data set contains the same information as DAT1 (except that DAT2 contains one record for each person). To transform DAT1 to DAT2 (or vice versa), you will need to use a two-dimensional array.

DAT1:

	ID	G1	G2	G3
1	1	A	B	F
2	1	B	A	C
3	2	B	A	D
4	2	C	B	C

DAT2:

	ID	M_G1	M_G2	M_G3	F_G1	F_G2	F_G3
1	1	A	B	F	B	A	C
2	2	B	A	D	C	B	C

Program 6.11 transforms data set DAT1 to data set DAT2. Because you are only creating the observation after you finish reading all the observations for each person, you need to use ID as the BY variable. Output is generated when LAST.ID equals 1. A one-dimensional array, G[3], is used to group the existing variables (G1–G3) from the input data set. A two-dimensional array, ALL_G[2,3], is used to create variables M_G1, M_G2, M_G3, F_G1, F_G2, and F_G3. The first dimension for ALL_G[2,3] is 2, which corresponds to the number of observations for each subject. The second dimension for ALL_G[2,3] is 3, which corresponds to the number of variables (G1–G3) that need to be transformed from the input data set. The SUM statement is used to increment index I in the implicit loop of the DATA step. Also, the iterative DO loop starts over with J = 1 when the implicit loop of the DATA step is executed. The newly created variables in the ALL_G array need to be RETAINED because to completely fill the array requires two (not one) iterations of the implicit loop of the DATA step. Otherwise the first three values in the array would always be reset to missing when all six are written out in the second iteration when LAST.ID is true (1). (*Program 6.11 is also illustrated in program6.11.pdf.*)

Program 6.11:

```
proc sort data = dat1;
     by id;
run;

data dat2(drop = i j g1 - g3);
    set dat1;
```

```
    by id;
    array all_g[2,3] $ m_g1 - m_g3 f_g1 - f_g3;
    array g[3];
    retain all_g;
    if first.id then i = 0;
    i + 1;
    do j = 1 to 3;
        all_g[i,j] = g[j];
    end;
    if last.id;
run;
```

Program 6.12 transforms data set DAT2 back to data set DAT1. The thinking behind Program 6.12 is similar to Program 6.11. A one-dimensional array, G[3], is used to create variables G1–G3. A two-dimensional array, ALL_G[2,3], is used to group the existing variables M_G1, M_G2, M_G3, F_G1, F_G2, and F_G3. The nested loop is used to create variables G1–G3. The OUTPUT statement that follows the inner loop is needed to create two observations per iteration of the outer loop.

Program 6.12:

```
data dat1(drop = i j m_g1 - f_g3);
    set dat2;
    array all_g[2,3] m_g1 - f_g3;
    array g[3] $;
    do i = 1 to 2;
      do j = 1 to 3;
          g[j] = all_g[i,j];
      end;
      output;/* After the inner J loop and before the end of the
                outer I loop */
    end;
run;
```

Exercises

Exercise 6.1. Consider the following data set, PROB6_1.SAS7BDAT:

	ID	G1	G2	G3	S1	S2	S3
1	1	A	A	C	3	4	9
2	2	A	B	F	3	7	4
3	3	A	C	B	5	8	9

Transform PROB6_1 to multiple observations per subject by using array processing like below:

	ID	TIME	GRADE	SCORE
1	1	1	A	3
2	1	2	A	4
3	1	3	C	9
4	2	1	A	3
5	2	2	B	7
6	2	3	F	4
7	3	1	A	5
8	3	2	C	8
9	3	3	B	9

Exercise 6.2. The PROB6_2.sas7bdat data set contains answers for a multiple-choice test for 1,000 students. There are 100 questions for this test (Q1–Q100). In addition to the Q1–Q100 variables, there is an ID variable in this data set. The first observation is the solution of the test. The next 1,000 observations are the answers from the students. Based on this data set, create a variable, SCORE, that is the number of correct answers for each student. For example, if student S1 has 82 correct answers compared to the KEY value, the SCORE will equal 82 for S1.

Exercise 6.3. Perry Watts's candidate exercise. Program 6.6 creates ABOVE1–ABOVE6 by comparing whether each measurement of SBP from the SBP2 data set is above (1) or below (0) the average SBP measurement of pre-treatment (140) and post-treatment measurements (120). Write a program by using array processing to calculate the average SBP values for pre-treatment (SBP1–SBP3) and post-treatment (SBP4–SBP 6) values for the four patients in the SBP2 data set.

7

Combining Data Sets

7.1 Vertically Combining Data Sets

When the data that you work with come from different sources, you need to combine the data before you generate your report(s) or perform statistical analysis. In a situation where observations are from a different source, the data need to be combined vertically. On the other hand, when variables are from a different source, the data need to be combined horizontally. In SAS®, two methods of combining data are concatenating and interleaving data.

7.1.1 Concatenating Data Sets

Concatenating data sets refers to creating a single data set by combining two or more data sets, one after another. The syntax for concatenating data sets is as follows:

```
SET data-set(s);
```

When concatenating data sets, SAS reads all the observations from the first data set and then continues to read all the observations from the second data set, and so on. The number of observations in the combined data set is the sum of the observations from the input data sets. If the input data sets have different variables, the observations from a given data set are set to missing for variables that exist only in other data sets. For example, Program 7.1 concatenates RECORD1 and RECORD2 data sets.

RECORD1:

	Name	Course	Score
1	John	MATH	90
2	John	MATH	85
3	Mary	MATH	.
4	Tom	MATH	92

RECORD2:

	Name	Course	Grade
1	Joe	ENGLISH	96
2	John	ENGLISH	89
3	Mary	ENGLISH	78
4	Tom	ENGLISH	.
5	Dave	ENGLISH	98

Program 7.1:

```
data ex7_1;
    set record1 record2;
run;

title 'Concatenating record1 and record2';
proc print data = ex7_1;
run;
```

Output from Program 7.1:

```
          Concatenating record1 and record2
      Obs  Name   Course   Score    Grade
       1   John    MATH      90        .
       2   John    MATH      85        .
       3   Mary    MATH       .        .
       4   Tom     MATH      92        .
       5   Joe     ENGL       .       96
       6   John    ENGL       .       89
       7   Mary    ENGL       .       78
       8   Tom     ENGL       .        .
       9   Dave    ENGL       .       98
```

Log from Program 7.1:

```
629 data ex7_1;
630    set record1 record2;
631 run;
```

```
WARNING: Multiple lengths were specified for the variable Course
        by input data set(s). This may cause truncation of data.
NOTE: There were 4 observations read from the data set
      WORK.RECORD1.
NOTE: There were 5 observations read from the data set
      WORK.RECORD2.
NOTE: The data set WORK.EX7_1 has 9 observations and 4
      variables.
NOTE: DATA statement used (Total process time):
      real time       0.01 seconds
      cpu time        0.00 seconds
```

When concatenating data sets, if common variables from different input data sets have a different data *type*, SAS will stop processing and issue an error message in the log. However, when common variables have different *length*, *label*, *format*, or *informat* attributes, SAS will use the attributes from the first data set that contains the variable with that attribute and issue a warning message in the log. In Program 7.1, because the length of the COURSE in the RECORD1 data set is 4, the length of the variable COURSE is also 4 in the combined data set. Thus, "ENGLISH" is truncated to "ENGL".

During the compilation phase when data sets are concatenated, the program data vector (PDV) is created based on the descriptor information for all the data sets in the SET statement. The contents in the PDV contain all variables from each of the input data sets, as well as variables, if any, that are being created in the DATA step.

During the execution phase, SAS reads each observation from the first data set into the PDV, executes additional statements if there are any in the DATA step, and finally writes the contents from the PDV to the output data set. As expected with a SET statement, variables from the input data set are *not* set to missing each time the implicit loop of the DATA step iterates. Also as expected, when new variables are created, they will be set to missing per iteration of the DATA step. However, because multiple data sets are being concatenated, an extra step occurs when SAS reaches the end-of-file marker in the first data set. At that time, all the values in the PDV (except for automatic variables) are set to missing. Then SAS begins the same process with observations from the second data set, and so on.

Because the variables GRADE and SCORE are used separately to record testing scores for English and Math, you can create a single SCORE variable from them with the RENAME= data set option:

RENAME= (old-name-1=new-name-1 <...old-name-n=new-name-n>)

In the RENAME= data set option, *old-name* is the variable you want to rename, and *new-name* is the newly given name of the variable. Program 7.2 utilizes the RENAME= option to rename GRADE to SCORE in the RECORD2 data set. In addition, the variable COURSE is set to length of 7 by the LENGTH statement.

Program 7.2:

```
data ex7_2;
    length Course $ 7;
    set record1 record2(rename = (grade = score));
run;

title 'Renaming the GRADE variable before concatenating the
data';
proc print data = ex7_2;
run;
```

Output from Program 7.2:

```
Renaming the GRADE variable before concatenating the data
           Obs      Course        Name        Score
            1       MATH          John          90
            2       MATH          John          85
            3       MATH          Mary           .
            4       MATH          Tom           92
            5       ENGLISH       Joe           96
            6       ENGLISH       John          89
            7       ENGLISH       Mary          78
            8       ENGLISH       Tom            .
            9       ENGLISH       Dave          98
```

7.1.2 Interleaving Data Sets

Interleaving data sets utilizes BY-group processing with the SET statement to combine two or more data sets vertically:

```
SET data-set(s);
BY variable(s);
```

The number of observations in the resulting data set is the sum of the observations from all of the input data sets. The observations in the resulting data sets are arranged by the values of the BY variable(s). Within each BY group, the order of the observation is arranged by the order of the input data sets. Similar to concatenating data sets, if the input data sets have different variables, the observations from the data set are set to missing for variables that exist only in other data sets.

   Before interleaving data sets, the input data sets must be sorted by the same variable(s) that you use in the BY statement. Program 7.3 begins with sorting RECORD1 and RECORD2 data sets by the NAME variable and then interleaves the sorted data sets in the DATA step.

Program 7.3:

```
proc sort data = record1 out = record1_sort;
    by Name;
run;

proc sort data = record2 out = record2_sort;
    by Name;
run;

data ex7_3;
    length Course $ 7;
    set record1_sort record2_sort;
    by Name;
run;
```

```
title 'Interleaving record1 and record2';
proc print data = ex7_3;
run;
```

Output from Program 7.3:

```
          Interleaving record1 and record2
   Obs    Course    Name     Score     Grade
    1     ENGLISH   Dave       .         98
    2     ENGLISH   Joe        .         96
    3     MATH      John      90          .
    4     MATH      John      85          .
    5     ENGLISH   John       .         89
    6     MATH      Mary       .          .
    7     ENGLISH   Mary       .         78
    8     MATH      Tom       92          .
    9     ENGLISH   Tom        .          .
```

During the compilation phase when interleaving data sets, SAS creates FIRST.VARIABLE and LAST.VARIABLE in addition to the variables from the input data sets in the PDV.

During the execution phase, SAS reads all the observations from the first BY group from the data set that contains the observations in the first BY group. If the BY group appears in more than one data set, SAS will read the observations from the data sets in the order in which they are listed in the SET statement. The SET statement sets the variables in the PDV to missing each time SAS begins to read a new data set or when the BY group changes. This process continues until SAS has read all the observations from all the input data sets.

7.2 Horizontally Combining Data Sets

Combining data sets horizontally is required when *variables* come from different sources. Horizontal data merges are a common occurrence. For example, to obtain a complete record for a patient in a health-care study, demographic profiles, medical information, and survey questions often have to be combined. This section covers different methods of combining data sets horizontally, such as one-to-one reading, one-to-one merging, match-merging, and updating data sets.

7.2.1 One-to-One Reading

One-to-one reading utilizes multiple SET statements to combine observations from two or more input data sets independently, forming one observation that contains all of the variables from each contributing data set.

Observations are combined based on their relative position in each data set. The first observation in the first data set, for example, is combined with the first observation in the remaining data sets. The number of observations in the combined data set is equal to the number of observations in the smallest input data set. If the input data sets share common variables, only the values read in from the last data set are kept. Here's the syntax for reading two input data sets using one-to-one reading:

```
SET data-set-1;
SET data-set-2;
```

Program 7.4 utilizes one-to-one reading to combine data sets RECORD1 and RECORD2. Notice that the number of observations in the combined data set is 4, which is equal to the number of observations in the smallest input data set (RECORD1). Because variables NAME and COURSE appear in both RECORD1 and RECORD2, only the values for NAME and COURSE in RECORD2 are written to the output data set.

Program 7.4:

```
data ex7_4;
    set record1;
    set record2;
run;

title 'Use One-to-one reading to combine record1 and record2';
proc print data = ex7_4;
run;
```

Output from Program 7.4:

```
     Use One-to-one reading to combine record1 and record2
         Obs     Name    Course    Score    Grade
          1      Joe      ENGL       90        96
          2      John     ENGL       85        89
          3      Mary     ENGL        .        78
          4      Tom      ENGL       92         .
```

During the compilation phase, SAS reads the descriptor information of the data sets in all the SET statements and creates a PDV that contains all the variables from all the input data sets plus variables, if any, that are being created in the DATA step.

During the execution phase, SAS reads the first observation from the first data set into the PDV. Then the second SET statement reads the first observation from the second data set into the PDV. If, at this point, both data sets share common variables, associated values from the first data set will be overwritten by values from the second data set. After executing any additional SAS statements in the DATA step, SAS writes the contents from the PDV to the output data set.

At the beginning of the second iteration, variables in the PDV are automatically retained, except for the variables being created in the DATA step. SAS continues reading observations from one data set to another until it reaches the end-of-file marker of the data set that contains the smallest number of observations.

The output data set from Program 7.4 shows how one-to-one reading works. However, the results are incorrect because of the shared variables, NAME and COURSE. John's MATH *score* (90), for example, is combined with Joe's ENGL *grade* (96). Furthermore, the fifth observation for Dave in RECORD2 never makes it to the output.

Although one-to-one reading is not advised for combining the RECORD1 and RECORD2 data sets in Program 7.4, the merge-type works well when an IF statement is used in conjunction with a SET statement. For example, Program 7.5 merges the *mean* for SCORE calculated from RECORD1 with detail records for the *same* variable from the *same* data set. To create EX7_5, detail records for SCORE from RECORD1 are read in first. Then SAS uses an IF statement along with a second SET statement to obtain MEAN_SCORE from RECORD1_MEAN, which was created by exercising PROC MEANS beforehand. Using SET and IF statements together in this fashion ensures that SAS will not encounter an end-of-file marker that would abruptly terminate the data step. Consequently, the single MEAN_SCORE can be associated with each observation in RECORD1.

Program 7.5:

```
proc means data = record1 noprint;
    var score;
    output out = record1_mean (keep = mean_score)
    mean = mean_score;
run;

data ex7_5;
    set record1;
    if _n_ = 1 then set record1_mean;
run;

title 'Use One-to-one reading to merge the mean score with
record1';
proc print data = ex7_5;
run;
```

Output from Program 7.5:

```
     Use One-to-one reading to merge the mean score with record1
                                               mean_
     Obs    Name    Course    Score    score
      1     John     MATH       90       89
      2     John     MATH       85       89
      3     Mary     MATH        .       89
      4     Tom      MATH       92       89
```

7.2.2 One-to-One Merging

One-to-one merging uses the MERGE statement, without an accompanying BY statement, to read observations from two or more input data sets into one observation in a combined data set. The result from a one-to-one *merging* is similar to what you get with one-to-one *reading* except that one-to-one merging processes *all* observations in *all* data sets listed in the MERGE statement. The syntax for one-to-one merging is as follows:

MERGE data-set(s);

Program 7.6 uses one-to-one merging to combine RECORD1 and RECORD2. Other than the last observation for Dave in EX7_6, results match output generated in Program 7.4.

Program 7.6:

```
data ex7_6;
    merge record1 record2;
run;

title 'Use One-to-one merging to combine record1 and record2';
proc print data = ex7_6;
run;
```

Output from Program 7.6:

```
    Use One-to-one merging to combine record1 and record2
       Obs    Name    Course    Score    Grade
        1     Joe      ENGL       90       96
        2     John     ENGL       85       89
        3     Mary     ENGL        .       78
        4     Tom      ENGL       92        .
        5     Dave     ENGL        .       98
```

MERGE works similarly to SET in that SAS reads the descriptor information of the input data sets during the compilation phase. SAS then creates the PDV that contains all the variables from all the data sets, along with new variables created in the DATA step.

During the execution phase with MERGE, observations from each data set are processed in the same order as they were with one-to-one reading. If two data sets have the same variables, the values from the second data set will replace the values that are read from the first data set. However, when SAS encounters the end-of-file marker from the data set with the fewest number of observations, variables in all the data sets involved with the merge will be set to missing in the PDV. Then SAS continues to read in data sets that have more observations until all observations from all data sets have been processed.

7.2.3 Match-Merging

Match-merging combines observations from two or more SAS data sets into a single observation according to the values of one or more common variables. The number of observations in the combined data set equals the sum of the largest number of observations in each BY group among all the input data sets. If the input data sets contain common variables that are not used as BY variables, values from the initial data sets listed in the MERGE statement will be overwritten by values contained in the last data set in the list when the observations being processed share the same BY values. To perform match-merging, you need to use the MERGE statement with the BY statement in the DATA step:

```
MERGE data-set(s);
BY variable(s);
```

All the input data sets must be previously sorted by the variable(s) listed in the BY statement.

The data sets being combined through match-merging can be related by one-to-one, one-to-many, or many-to-many matching of the values of one or more variables. One-to-one matching refers to a single observation in one data set relating to a single observation from another based on the value of one or more common variables. In the case of two or more common variables, one-to-one matching means that each combination of values occurs only once in each data set. One-to-many matching means a single observation in one data set is associated with multiple observations from another based on the value of one or more common variables. Many-to-many matching refers to multiple observations from each input data set can be related based on values of one or more common variables. Readers should be cautious in performing many-to-many match-merging. The MERGE statement does not produce a Cartesian product (all possible combinations of observations between the input data sets) on a many-to-many match-merge. In order to get a Cartesian product, you can use ("a" or "the") SQL procedure, which is not covered in this book.

Program 7.7 merges RECORD1_SORT and RECORD2_SORT by the common variable NAME. Both data sets have been previously sorted by NAME in Program 7.3. The variable COURSE is also a common variable in both input data sets; however, COURSE is not used as a BY variable. Thus, the values read in from RECORD2_SORT replace the values read from RECORD1_SORT. As an aside, recall that "ENGLISH" is truncated to "ENGL" with 4 bytes, because "MATH" precedes "ENGLISH" in the input data.

Notice that "John" has two observations in RECORD1_SORT but only one observation in RECORD2_SORT. Only the first value for COURSE for subject "John" is replaced with "ENGL" in WORK.EX7_7, but the second value for COURSE carries John's "MATH" observation over from RECORD1_SORT. Because SCORE is found only in RECORD1_SORT, both values (90 and 85) in the combined data set for "John" originate in RECORD1_SORT. On the other hand, GRADE exists only in RECORD2_SORT, but because there is only one observation for "John" in RECORD2_SORT, the value for the GRADE variable (89) is carried down to the second observation of "John". Data sets RECORD1_SORT and RECORD2_SORT are displayed below so that you can follow along in this discussion.

RECORD1_SORT:

	Name	Course	Score
1	John	MATH	90
2	John	MATH	85
3	Mary	MATH	.
4	Tom	MATH	92

RECORD2_SORT:

	Name	Course	Grade
1	Dave	ENGLISH	98
2	Joe	ENGLISH	96
3	John	ENGLISH	89
4	Mary	ENGLISH	78
5	Tom	ENGLISH	.

Program 7.7:

```
data ex7_7;
    merge record1_sort record2_sort;
    by Name;
run;

title 'Use match-merge to combine record1 and record2';
proc print data = ex7_7;
run;
```

Output from Program 7.7:

```
        Use match-merge to combine record1 and record2
        Obs    Name    Course    Score    Grade
         1     Dave     ENGL        .       98
         2     Joe      ENGL        .       96
         3     John     ENGL       90       89
         4     John     MATH       85       89
         5     Mary     ENGL        .       78
         6     Tom      ENGL       92        .
```

During the compilation phase, SAS reads the descriptor information of the input data sets listed in the MERGE statement. SAS then creates the PDV that contains all the variables from all the data sets along with new variables created in the DATA step. In addition, SAS creates FIRST. VARIABLE and LAST.VARIABLE for each variable listed in the BY statement in the PDV.

During the execution phase, SAS examines the first BY group in each input data set listed in the MERGE statement to determine which BY group should appear first in the output data set. Then the DATA step starts to read observations from the first BY group from each data set. When reading the first BY group, the DATA step reads the first observations into the PDV in the order in which they appear in the MERGE statement. If a data set does not have observations in that BY group, the PDV contains missing values for the variables unique to that data set. After reading the first observation from the last data set, SAS executes other statements in the DATA step (if there are any). At the end of the DATA step, SAS writes the contents of the PDV to the output data set and returns to the beginning of the next iteration. SAS retains the values of all variables in the PDV except those variables that were created by the DATA step; SAS sets those values to missing. SAS continues to merge observations until it writes all observations from the first BY group to the output data set. When SAS has read all observations in a BY group from all data sets, it sets all variables in the PDV (except those created by SAS) to missing.

After reading all the observations from all the data sets from the first BY group, SAS begins to examine the next BY group in each data set to determine which BY group should appear next in the new data set. SAS starts to process the next BY group in each data set. The merging process continues until the DATA step reads all observations from all BY groups in all data sets.

Program 7.7 was not a logical approach to merge RECORD1 and RECORD2 because the COURSE variable does not convey any meaningful information for the combined data set. Program 7.8 improves on Program 7.7 by utilizing the DROP= and RENAME= data set options.

Program 7.8:

```
data ex7_8;
    merge record1_sort(drop = course
                       rename = (score = Math_score))
          record2_sort(drop = course
                       rename = (grade = English_score)) ;
    by Name;
run;

title 'An improved approach to merge record1 and record2';
proc print data = ex7_8;
run;
```

Output from Program 7.8:

```
        An improved approach to merge record1 and record2
                          Math_    English_
            Obs     Name    score     score
             1      Dave       .        98
             2      Joe        .        96
             3      John      90        89
             4      John      85        89
             5      Mary       .        78
             6      Tom       92         .
```

By default, SAS combines all the observations from all the input data sets during a match-merge. You can also exclude any unmatched observations by using the IN= data set option, which has the following form:

`IN=variable`

The *variable* in the IN= data set option is a temporary variable that is available in the PDV but is not included in the output data set. The value of the *variable* is either 1 or 0. The *variable* equals 1 if the input data set contributes to the current observation in the PDV; otherwise, its value equals 0. In addition to the MERGE statement, the IN= data set option can be used with the SET, MODIFY, and UPDATE statements.

Program 7.9 excludes the unmatched observations that were read from RECORD1_SORT and RECORD2_SORT. Two temporary variables, IN_RECORD1 and IN_RECORD2, are created by using the IN= data set option. IN_RECORD1 equaling 1 indicates the current observation in the PDV is read from the RECORD1_SORT data set. Similarly, IN_RECORD2 equaling 1 indicates the current observation in the PDV is read from RECORD2_SORT. A subsetting IF statement is used to include observations that are read from both RECORD1_SORT and RECORD2_SORT data sets.

Program 7.9:

```
data ex7_9;
    merge record1_sort (drop = course
                        rename = (score = Math_score)
                        in = in_record1)
          record2_sort (drop = course
                        rename = (grade = English_score)
                        in = in_record2);
    by Name;
    if in_record1 and in_record2;/*A SUBSETTING IF
    STATEMENT */
run;
```

```
title 'Excluding unmatched observations';
proc print data = ex7_9;
run;
```

Output from Program 7.9:

```
            Excluding unmatched observations
                           Math_    English_
        Obs    Name        score     score
         1     John          90        89
         2     John          85        89
         3     Mary           .        78
         4     Tom           92         .
```

7.2.4 Updating Data Sets

You can use the UPDATE statement to update a *master* data set with a *transaction* data set. The *master* data set refers to the data set that contains the original information. The *transaction* data set refers to the data set that contains newly collected information. The UPDATE statement uses observations from the *transaction* data set to change the values of the matched observations in the *master* data set. The number of observations in the resulting data set is the sum of the observations in the *master* data set and number of unmatched observations in the *transaction* data set. To update a data set, you must use the BY statement after the UPDATE statement. The syntax for updating data sets is as follows:

```
UPDATE master-data-set transaction-data-set;
BY variable(s);
```

The *master-data-set* in the UPDATE statement names the SAS data set that contains the original information, while *transaction-data-set* is the name of the SAS data set that contains new information. When the *transaction* data set contains duplicate values of the BY variable, only the last values copied to the PDV are written to the output data set. If the *master* data set contains duplicate values of the BY variable, only the first observation in the *master* data set is updated and SAS will write a warning message in the log.

Updating data sets is similar to match-merging with the MERGE statement. However, unlike MERGE, missing values in the *transaction* data set do not replace the existing values in the *master* data set. Thus, if you would like to update only some observations but not all observations of a specific variable, you can set the observations you do not want to change to missing in the *transaction* data set.

For example, some of the test scores in the RECORD2_SORT data set need to be modified. In Program 7.10, the *transaction* data set, RECORD3_ SORT, contains newly collected scores that are stored in the GRADE variable. Program 7.10 updates RECORD2_SORT with values from GRADE in RECORD3_SORT. Notice that the values for GRADE for Joe and John are missing in RECORD3_SORT. Thus, in the final updated data set, the values for Joe and John originate in the *master* data set (RECORD2_SORT).

RECORD3:

	Name	Grade
1	Joe	.
2	John	.
3	Mary	82
4	Tom	90
5	Dave	97

Program 7.10:

```
proc sort data = record3 out = record3_sort;
    by Name;
run;

data ex7_10;
    update record2_sort record3_sort;
    by name;
run;

title 'Updating record2 data set';
proc print data = ex7_10;
run;
```

Output from Program 7.10:

```
Updating record2 data set
Obs   Name    Course     Grade
 1    Dave    ENGLISH     97
 2    Joe     ENGLISH     96
 3    John    ENGLISH     89
 4    Mary    ENGLISH     82
 5    Tom     ENGLISH     90
```

Exercises

Exercise 7.1. Consider the following data set, GROUP_SCORE.SAS7BDAT. Use one-to-one reading to add one variable, MAXSCORE, that contains the maximum score of the SCORE variable.

	Group	Score
1	A	6
2	A	9
3	B	2
4	A	8
5	A	8
6	A	9
7	A	6
8	A	6
9	B	1
10	B	2

Exercise 7.2. Based on the same data set in Exercise 7.1, GROUP_SCORE. SAS7BDAT, use the match-merge method to add one variable, MEANSCORE, that contains the mean of the SCORE variable for each group.

8

Data Input and Output

8.1 Introduction to Reading and Writing Text Files

A SAS® data set is often created by reading data directly from an external text file. Observations (rows) and variables (columns) in the SAS data set correspond to records and the fields in the text file. You can go in the opposite direction and create a text file from a SAS data set. This section provides an overview for reading and writing text files, as well as an introduction to SAS informats and formats.

8.1.1 Steps for Reading Text Files

To read an external text file into SAS, you need to use an INFILE statement followed by the INPUT statement. The INFILE statement has the following form:

INFILE file-specification <options>;

File-specification is used to specify the location of the input data set. It can take the form of *'external-file'* (with quotes) or *fileref* created with a FILENAME statement. The simple form of the FILENAME statement is as follows:

FILENAME fileref 'external-file'

The FILENAME statement associates fileref with the external file. The following code shows how INFILE and FILENAME work together:

```
filename ex 'c:\example.txt';
data sasdat;
    infile ex;
    input...;
run;
```

If you don't want to use INFILE with FILENAME, the following is also perfectly acceptable:

```
data sasdat;
    infile 'c:\example.txt';
    input...;
run;
```

In addition, if you use the DATALINES or DATALINES4 statements, the INFILE statement can be omitted when options are not specified. To specify options in the INFILE statement with DATALINES or DATALINES4, just enter either *file-specification* as

INFILE DATALINES | DATALINES4 <options>;

Once the location of the external file is identified by the INFILE statement, you need to use the INPUT statement to read the data.

SAS provides many methods of reading external text files, which include the *column, formatted, list,* and *named* input methods. These input methods can also be mixed when reading an external file. In addition to these input methods, many options in the INFILE statement assist in importing the data. Choosing the correct input method and the option in the INFILE statement depends upon how the data is arranged or formatted in the text file. Thus, before reading a text file into SAS, one needs to examine the data carefully and then choose the correct method based on how data are stored in the text.

There are many aspects of the data that you need to inspect. Here is a checklist that should help you out:

- Are the data arranged in a fixed field or free format?
- Is the length of each record the same or varied?
- Does any numeric data field contain any nonstandard numeric values?
- Are delimiters used for separating each field?
- Is the maximum length of character data field greater than 8 bytes?
- Does the character data field contain any embedded blanks?
- Are the missing values located in the beginning, middle, or end of the record?
- Are you creating one observation based on multiple records or are you creating multiple observations based on one record?

The text files shown in this chapter contain data formats that represent the most commonly encountered scenarios. Some rarely seen data formats will not be presented. For example, the *named* input method is not presented. You can use the *named* input method to read a data file that contains values preceded by the name of the variable and an equal sign (=).

8.1.2 Steps for Writing Text Files

Creating a text file from a SAS data set often starts with the FILE statement and is followed by the PUT statement in the DATA step. You use the FILE statement to specify the name of the output file that you want to create, which has the following form:

FILE file-specification <options>;

If *file-specification* equals an *'external file'* or a *fileref,* output from the PUT statement (introduced in Chapter 3) is written to an external file. On the other hand, if you specify PRINT as the *file-specification,* SAS will redirect the PUT output to the output window. As with INPUT, SAS provides four types of PUT statements: *column, formatted, list,* and *named.* Again, the *named* output method, like the *named* input method, is not covered in this chapter.

8.1.3 Data Informat

An informat tells SAS how to read a field from a text file. Its use depends upon the type of INPUT statement being used. The SAS informat takes the following general form:

<$>informat<w>.<d>

The optional dollar sign ($) is used for reading character data. The *informat* component is the name of the informat. The w value is used to specify the width (or the number of columns) of the input field. The d value is used for dividing the input numeric data by 10 to the power of d.

The w. informat is for reading standard character data values. When using the w. informat, SAS trims the leading blanks of the character value and left-aligns the values. The w. informat treats blank fields or a single period (.) as missing and converts the period (.) to a blank when the missing value is stored. If, however, the w. informat encounters two consecutive periods in the external file, the field is treated as a non-missing character value.

SAS also provides a wide variety of informats for reading other types of character data. For example, if you do not want to trim leading blanks or do not want to treat a single period as a missing value when reading in character values, you can use the *$CHARw.* informat. Or if you want to convert all the character data to uppercase, you can use the *$UPCASEw.* informat. Using the *$QUOTEw.* informat removes matching quotation marks from character data, and so on.

In SAS, standard numeric values can contain only numbers, decimal points, numbers in scientific notations, and plus or minus signs. Nonstandard numeric values can consist of fractions, values that contain special characters (such as the percent sign, %), the dollar sign ($), commas (,), etc.

TABLE 8.1

Examples of Using *w.d* Informat

Raw Data Value	Informat	Value That Read into SAS	Note
3.14159	7.	3.14159	
3.14159	6.	3.1415	Only the first 6 fields are read
3.14159	7.5	3.14159	5 in the informat is ignored
314159	6.5	3.14159	5 is used to specify 5 decimal points

The *w.d* informat is used for reading standard numeric data. When using the *w.d* informat, the *d* value will be ignored if the input data contains decimal points. The *w.d* informat can handle scientific notation and will interpret a single period as a missing value. Table 8.1 shows by example how the *w.d* informat works.

A commonly used informat for reading nonstandard numeric values is the *COMMAw.d* informat. The *COMMAw.d* informat removes embedded characters, including commas, blanks, dollar signs ($), percent signs (%), dashes (-), and closed parentheses, from an input data field. The *COMMAw.d* informat also converts the open parenthesis at the beginning of a field to a minus sign.

8.1.4 Data Format

A format tells SAS how to write a SAS data set value out to a text file. The general form of a SAS format is as follows:

```
<$>format<w>.<d>
```

The dollar sign ($) is used to indicate a character format. The *format* is the name of the format, which can be either a SAS format or a user-defined format. The *w* value is the format width, which is the number of columns in the output data in most cases. The *d* value is used to indicate the decimal scaling factor in the numeric formats. A format always contains a period (.) as a part of the format name.

Chapter 1 introduced some of the commonly used formats, such as the standard character format ($*w*.), standard numeric format (*w.d*), and a few nonstandard character and numeric formats. Readers can refer to SAS documentation for other types of formats.

8.1.5 SAS Date and Time Values

In SAS, date or time values are stored as numeric values. A SAS date value, which accounts for all leap-year days, is a value that represents the number of days between January 1, 1960, and a specified date. Dates after

January 1, 1960, are positive numbers, and dates before January 1, 1960, are negative numbers. Because dates are stored as numbers, you can easily perform any kind of computation on dates, such as calculating the number of days between two specified dates.

SAS time value is a value that stands for the number of seconds since midnight of the current day and ranges between 0 and 86,400. SAS datetime values correspond to the number of seconds between midnight January 1, 1960, and a specified second within a specified date.

SAS provides a large selection of informats to read different notations of date and time and stores the data to a SAS date, time, or datetime value. Date and time values can be read into SAS successfully provided that the date and time are separated by either a blank or the following delimiters: ! # $% & () * + −./: ; < = > ? [\] ^ _ {|} ~. However, only one type of delimiter can be used; for example, 10/01/49 is a valid input date value, while 10-01/49 is not.

The *DATEw.*informat is used for reading in date values in the form of *ddmmmyy* or *ddmmmyyyy*. The *dd* component is an integer ranging from 01 to 31 that denotes the day of the month, whereas *mmm* is the first three letters of the month. The *yy* or *yyyy* component is the year in either a two-digit or four-digit format. Each of the three components can be separated by blanks or the delimiter that was listed in the previous paragraph. The *MMDDYYw.* informat is similar to the *DATEw.* informat except that the *MM* value, representing month, is an integer from 01 to 12.

Examples of date informats are listed in Table 8.2. Readers can refer to SAS documentation for other types of date and time informats.

A SAS date value can contain either a two-digit or four-digit year component. SAS interprets the two-digit year based on a 100-year span that starts with the value specified in the YEARCUTOFF = option. By default, the YEARCUTOFF = option is set to the year 1920, meaning that the 100-year span is from 1920 to 2019.

Thus, a two-digit value of 49 will be interpreted as 1949. If you set the YEARCUTOFF = option to 1950, then the 100-year span is from 1950 to 2049. Thus, 49 will be treated as 2049. To avoid confusion, you should try to avoid using a two-digit year option if possible. Also, if your data spans more than 100 years, you must use the four-digit year option.

TABLE 8.2

Examples of Using Date Informats

Date Input	Informat	Result
01oct1949	date9.	-3744
01oct49	date7.	-3744
01-oct-1949	date11.	-3744
10-01-49	mmddyy8.	-3744
10/01/1949	mmddyy10.	-3744
01 10 1949	ddmmyy10.	-3744

TABLE 8.3

Examples of Date Formats

Numeric Value	Format	Formatted Value
-3744	date5.	01OCT
-3744	date7.	01OCT49
-3744	date9.	01OCT1949
-3744	date11.	01-OCT-1949
-3744	mmddyy6.	100149
-3744	mmddyy8.	10/01/49
-3744	mmddyy10.	10/01/1949
-3744	mmddyyp10.	10.01.1949

Dates and times are stored as numbers in SAS, so you need to use a format to see what the numbers represent. For example, *DATEw.* writes date values in the form of *ddmmmyy, ddmmmyyyy,* or *dd-mmm-yyyy,* where *dd* is an integer that represents the day of the month, *mmm* is the first three letters of the month, and *yy* or *yyyy* is a two-digit or four-digit integer that represents the year. The *w* value represents the width of the output field.

Another commonly used date format is the *MMDDYYw.* format that writes date values in the form of *mmdd<yy>yy* or *mm/dd/<yy>yy,* where *mm* is an integer value that represents month, *dd* is an integer value that represents day, and *yy* or *yyyy* is a two-digit or four-digit integer that represents the year. The forward slash is a separator between month, day, and year.

The *MMDDYYxw.* format is similar to *MMDDYYw.* except that you can use the following *x* field to specify a separator: B (a blank), C (a colon), D (a dash), N (no separator), P (a period), and S (a slash). Some examples of date formats are listed in Table 8.3.

You can use the *DATETIMEw.d* format to write datetime values in the form of *ddmmmyy:hh:mm:ss.ss.* The *hh* value is an integer that represents the hour in 24-hour clock time, *mm* is an integer that represents the minutes, and *ss.ss* is the number of seconds to two decimal places. The *w* value is used to specify the width of the output field, and the optional *d* value is used to indicate the number of digits to the right of the decimal point in seconds value.

8.2 Reading Text Files

The INPUT statement is used for reading the external text file into SAS. The INPUT statement first copies the data from the external text file to the input buffer and then controls how values are transferred from the input buffer into the program data vector (PDV). The DATA step execution for reading a text file is illustrated in Chapter 3.

8.2.1 Column Input

The column input method was discussed in Chapter 1 and again in Chapter 3. You can use the column input method only to read a text file that contains variables in a fixed field with character or standard numeric values. The syntax for reading a text file using the column input method is as follows:

```
INPUT variable <$> start-column <- end-column>
```

During the execution phase of the DATA step, each record from the text file is read into the input buffer. SAS uses the input pointer to read data from the input buffer to the PDV that follows the instruction from the INPUT statement, which then creates observations for the SAS data set. The compilation and execution phases for reading an external text file via the column input method were shown in Program 3.1 in Chapter 3.

Using the column input method does not require that each field in the text file be separated by a delimiter. The character values in the text file can contain embedded blanks. The input values in the text file can also be read in any order. Any blank fields in the text file are treated as missing by the INPUT statement. In addition, the INPUT statement treats a single period in either numeric or character fields as missing.

A text file that contains records of varied length will cause problems when the column input method is used. For example, CH8 1.TXT contains data in a fixed-field format. However, the length of each record varies, because the last variable CITY is left-aligned, and CITY names have different lengths. A description of the data is presented in Table 8.4.

By default, SAS uses the FLOWOVER option in the INFILE statement. That is to say, if a record is shorter than expected, the INPUT statement will read the values from the following data record. To read the data correctly, you can use the TRUNCOVER option in the INFILE statement. TRUNCOVER accommodates variable-length records.

Alternatively, you can use the PAD option in the INFILE statement. The PAD option pads each record with blanks so that all the records have the same length. Using either the TRUNCOVER or the PAD option is especially useful when you read records that contain missing values at the end. Program 8.1 uses the column input method along with the TRUNCOVER option to read CH8_1.TXT.

TABLE 8.4

Field Description for CH8_1.TXT

Variable Name	Locations	Variable Type
Name	Columns 1–14	Character
Gender	Column 16	Character
Age	Columns 17–18	Numeric
City	Columns 20–30	Character

CH8_1.txt:

```
12345678901234567890123456789 0
James Miller     M26 New York
Patricia Davis F30 Phoenix
John Jackson     M31 Dallas
Mary Martinez   F32 Los Angeles
```

Program 8.1:

```
data ex8_1;
    infile "W:\SAS_Book\dat\ch8_1.txt" truncover;
    input Name    $ 1 - 14
        Gender   $ 16
          Age      17 - 18
         City    $ 20 - 30;
run;

title 'Using the column input method and the TRUNCOVER option';
proc print data = ex8_1;
run;
```

Output from Program 8.1:

```
Using the column input method and the TRUNCOVER option
Obs          Name          Gender   Age      City
 1      James Miller          M       26     New York
 2      Patricia Davis        F       30     Phoenix
 3      John Jackson          M       31     Dallas
 4      Mary Martinez         F       32     Los Angeles
```

8.2.2 Formatted Input

If the input data set in fixed field contains nonstandard numeric values, you can use the formatted input method to read data. Here is the syntax of the formatted input method:

```
INPUT <pointer-control> variable informat;
```

The *pointer-control* can be either column or line pointer-control. Column pointer-control is used to move the input pointer in the input buffer to a specified column. Here are two commonly used column pointer-controls:

```
@n
+n
```

By default, the column input pointer is located at column 1 in the input buffer. The @*n* pointer-control moves the input pointer to column *n* in the input buffer, while the +*n* moves the pointer *n* columns forward to a column relative to the current position.

TABLE 8.5

Field Description for CH8_2.TXT

Variable Name	Locations
Name	Columns 1–14
Ethnic	Column 16
DOB	Columns 18–27
Income	Columns 30–39

For example, data in CH8_2.TXT are in fixed field with the last two fields containing nonstandard numeric values (description of the data is presented in Table 8.5). Program 8.2 reads CH8_2.TXT by using the formatted input method. The PAD option is also used because the length of each record is not the same. In the INPUT statement, the @1pointer-control sets the input pointer at column 1. Because the default position for the input pointer is at column 1, @1 can be omitted. The informat $14. instructs SAS to copy the value from the input buffer to the variable NAME in the PDV. Next, the input pointer is located at column 15. The +1 pointer-control moves the input pointer to column 16. The $1. informat then creates the ETHNIC variable. The pointer-control @18 positions the input pointer to column 18, and the *MMDDYY10.* informat instructs SAS to read values for the DOB variable. The last pointer-control +2, along with the *COMMA10.* informat, reads the value for the INCOME variable:

CH8_2.txt:

```
12345678901234567890123456789012345678990
James Miller    W 03-02-1986    $1,100,000
Patricia Davis B 11-13-1982    $90,000
John Jackson    W 06-15-1981    $75,000
Mary Martinez   H 08-02-1980    $150,000
```

Program 8.2:

```
data ex8_2;
    infile "W:\SAS_Book\dat\ch8_2.txt" pad;
    input @1 Name $14.
          +1 Ethnic $1.
          @18 DOB mmddyy10.
          +2 Income comma10.;
run;

title 'Using the formatted input method and the PAD option';
proc print data = ex8_2;
run;

proc contents data = ex8_2 varnum;
run;
```

Output from Program 8.2:

```
Using the formatted input method and the PAD option
   Obs        Name          Ethnic    DOB     Income
    1      James Miller        W      9557    1100000
    2      Patricia Davis      B      8352      90000
    3      John Jackson        W      7836      75000
    4      Mary Martinez       H      7519     150000
```

Partial output from PROC CONTENTS:

```
          Variables in Creation Order
          #    Variable    Type    Len
          1    Name        Char     14
          2    Ethnic      Char      1
          3    DOB         Num       8
          4    Income      Num       8
```

Program 8.2 also contains a CONTENTS procedure. Based on the partial output from PROC CONTENTS, the lengths for the NAME and ETHNIC variables are 14 and 1, respectively, which have the same w values in the character informats ($14. and $1.). However, both lengths for the DOB and INCOME variables are 8, which is different from the w values in the MMDDYY10. and COMMA10. informats. When using a formatted input method, the informat for the character variable is used to specify the number of columns to read in the input buffer; it is also used to specify the length of the variable. However, this rule does not apply to numeric variables. By default, SAS stores numeric values as floating-point numbers in 8 bytes, which is the default length for numeric variables. The w value in the informat for the numeric value does not affect the default length of the numeric variable but only serves to specify the number of columns to read.

8.2.3 List Input

The list input method is used to read free format data. Here is the syntax of the list input method:

```
INPUT variable <$>;
```

The optional dollar sign ($) in the INPUT statement is used to read character variables. To use the list input method, data in all the fields must contain either character values without any embedded blanks or standard numeric values. Each field needs to be separated by at least one blank or some other type of delimiter. The INPUT statement reads all the fields in each record from left to right, and you cannot skip or re-read any fields from the input data.

CH8_3.TXT contains data in free format with each field separated by a blank. The first and second fields contain first and last name, followed by age, number of people in the household, and the number of children in the household. Program 8.3 reads the CH8_3.TXT file by using the list input method.

CH8_3.txt:

```
12345678901234567890123456789 0
James Miller    26 3 1
Patricia Davis 30 4 1
John Jackson    31 2 0
Mary Martinez   32 4 2
```

Program 8.3:

```
data ex8_3;
    infile "W:\SAS_Book\dat\ch8_3.txt";
    input fname $
          lname $
          age
          family
          children;
run;

title 'Using the list input method';
proc print data = ex8_3;
run;
```

Output from Program 8.3:

```
                Using the list input method
    Obs     fname       lname      age   family   children
     1      James       Miller      26     3         1
     2      Patricia    Davis       30     4         1
     3      John        Jackson     31     2         0
     4      Mary        Martinez    32     4         2
```

The data execution for reading data by using the list input is different from the column input method. Instead of searching for specific columns, SAS scans the input buffer from left to right to copy the values from the input buffer to the PDV. When the INPUT statement begins to execute, the input pointer is located at column 1 and then reads the first field until it encounters a blank space. The blank space is used to indicate the end of the field. Once the first field is copied from the input buffer to the PDV, SAS starts to scan the input buffer until the next nonblank column is found, and then the second value is read from the input buffer until another blank column is reached. This process continues until all the fields are read from the input buffer.

By default, all the character variables are assigned with default length of 8. Thus, character values longer than 8 will be truncated. To correctly read character values longer than 8, you can use the LENGTH statement to preset the variable length before reading the data.

The list input method can read missing values correctly if the missing value is represented as a period. If the missing value is represented as blanks and is located at the end of the record, you can use the MISSOVER option in the INFILE statement to read the missing value correctly.

The data in CH8_4.TXT are in free format. Notice that the fourth record has "Christopher" (11 characters) as the first name, and the last two fields of the fourth record are missing. Program 8.4 reads the data correctly by using the MISSOVER option and sets the length for FNAME to 11 by using the LENGTH statement.

CH8_4.txt:

```
12345678901234567890123456 7890
James Miller        26 3 1
Patricia Davis      30 4 1
John Jackson        31 2 0
Christopher Jones 25
Mary Martinez       32 4 2
```

Program 8.4:

```
data ex8_4;
    infile "W:\SAS_Book\dat\ch8_4.txt" missover;
    length fname $11.;
    input fname $
          lname $
          age
          family
          children;
run;

title 'Using the LENGTH statement and the MISSOVER option';
proc print data = ex8_4;
run;
```

Output from Program 8.4:

```
        Using the LENGTH statement and the MISSOVER option
    Obs     fname          lname        age    family    children
    1       James          Miller       26       3          1
    2       Patricia       Davis        30       4          1
    3       John           Jackson      31       2          0
    4       Christopher    Jones        25       .          .
    5       Mary           Martinez     32       4          2
```

When data are arranged in free format, each field can be separated by a delimiter. In this situation, you need to use the DLM= option in the INFILE

TABLE 8.6

Use of INFILE Options for Listing Input when the Text File Contains Delimiter(s) and Missing Values (Missing Values Are Represented as Blanks)

Delimiter	Missing Value Location	INFILE Statement Option
Nonblank	None	`DLM=`
blank(s)	End of the record	`MISSOVER`
comma (,)	End of the record	`DSD`
Other[a]	End of the record	`DSD and DLM=`
comma (,)	Beginning of the record	`DSD`
Other[a]	Beginning of the record	`DSD and DLM=`
comma (,)	Middle of the record	`DSD`
Other[a]	Middle of the record	`DSD and DLM=`

[a] Delimiters other than blanks and commas.

statement when reading the data. For example, suppose that commas (,) are used as delimiters to separate each file; you can write the following statement:

```
infile "W:\SAS_Book\dat\ch8_4.txt" DLM = ",";
```

If missing values are represented as blanks and exist in either the beginning or the middle of a record in a free-format data set, each field in the text file needs to be separated by a nonblank delimiter in order to be read into SAS correctly. Table 8.6 summarizes the use of options in the INFILE statement when text files contain delimiter(s) and missing values (represented as blanks).

When using the DSD (Delimiter-Sensitive-Data) option in the INFILE statement, SAS sets the delimiter to a comma, treats two consecutive delimiters as a missing value, and removes quotation marks from values. When the missing value exists at the beginning of the record, only one delimiter is needed.

CH8_5.TXT is similar to CH8_3.TXT except that a field for the new GROUP variable has been added to the beginning of each record. Each field in CH8_5.TXT is also separated by commas. Notice that there is a missing value at the beginning of the first record, the end of the second record, and the middle of the third record. In this situation, the DSD option added to the INFILE statement is used to read in the data. DSD works correctly without the DELIMITER = option because DSD expects a comma by default. (See Program 8.5.)

CH8_5.txt:

```
12345678901234567890123456789 0
,James,Miller,26,3,1
B,Patricia,Davis,30,4,
B,John,Jackson,,2,0
A,Mary,Martinez,32,4,2
```

Program 8.5:

```
data ex8_5;
    infile "W:\SAS_Book\dat\ch8_5.txt" dsd;
    input group $
          fname $
          lname $
          age
          family
          children;
run;

title 'Using the DSD option';
proc print data = ex8_5;
run;
```

Output from Program 8.5:

```
                    Using the DSD option
  Obs    group    fname      lname      age    family   children
   1              James      Miller      26       3         1
   2       B      Patricia   Davis       30       4         .
   3       B      John       Jackson      .       2         0
   4       A      Mary       Martinez    32       4         2
```

8.2.4 Modified List Input

The modified list input is a more flexible method that can read data in free format and can read character values containing embedded blanks and nonstandard numeric values. The general syntax of modified list input is as follows:

```
INPUT variable <:|&|~> <informat>;
```

You can use the colon (:) format modifier along with an informat to read nonstandard numeric values and character values with more than eight characters. However, the character values cannot contain any embedded blanks. When using the colon (:) format modifier, SAS reads values until it reaches a blank column, a defined length of a variable (character only), or the end of the data line, whichever comes first.

The ampersand (&) format modifier enables you to read character values with embedded blanks with or without specifying a character informat. If the character values have more than 8 bytes, you can either use the LENGTH statement to prespecify the length or insert the $w. informat after the ampersand (&).

The ampersand (&) format modifier works for values that have one or more nonconsecutive blanks, because processing continues to the next field only when two or more *consecutive* blanks are encountered in the text file.

The tilde (~) format modifier enables you to read and retain single or double quotation marks, as well as delimiters within character values. When using the tilde (~) format modifier, you must use the DSD option in the INFILE statement; otherwise, this option will be ignored by the INPUT statement. Please refer to SAS documentation for an example.

CH8_6.TXT contains data in free format. The first field contains values for first name, followed by last name, city, and income information. In this text file, the first name of the fourth record, "Christopher", has 11 characters. Some values in the third field (city) contain embedded blanks. The last field (income) contains nonstandard numeric values. Thus, a modified input method needs to be used to read CH8_6.TXT.

In Program 8.6, the colon (:) format modifier and the informat $11. are used to read the value for the FNAME variable. Because all values for the LNAME variables are less than or equal to 8, only the regular list input method is used for reading LNAME. Next, because the city field contains embedded blanks and some city values are longer than eight characters, the ampersand (&) format modifier and the $11. informat are used to read values for the CITY variable. The ampersand (&) format modifier tells SAS that the value for CITY should be read until two consecutive blanks are reached. In the end, the colon (:) format modifier and the *COMMAw.d* informat are used to read nonstandard numeric values for the INCOME variable.

The use of an informat in modified list input is different from that for formatted input. In a formatted input, the informat is used to determine the length of the variable and the number of columns to read in the text file. However, the informat in a modified list input is used only to determine the length of the variable. The list input reads values in each field until the next blank is encountered. In Program 8.6, the *w* value is not specified in the *COMMAw.d* informat because the default length for numeric variables is eight, and specifying *w* in the *COMMAw.d* informat is not necessary. However, you need to specify the *w* value in the *COMMAw.d* informat in the formatted input because the *w* value is used to determine the number of columns to be read.

CH8_6.txt:

```
12345678901234567890123456789012345678901234567890
James Miller New York            $1,100,000
Patricia Davis Phoenix           $90,000
John Jackson Dallas              $75,000
Christopher Jones San Francisco  $95,000
Mary Martinez Los Angeles        $150,000
```

Program 8.6:

```
data ex8_6;
    infile "W:\SAS_Book\dat\ch8_6.txt";
    input fname : $11.
```

```
        lname $
        city & $13.
        income : comma.;
run;

title 'Using the modified list input method';
proc print data = ex8_6;
run;
```

Output from Program 8.6:

```
            Using the modified list input method
    Obs    fname         lname      city              income
     1     James         Miller     New York         1100000
     2     Patricia      Davis      Phoenix            90000
     3     John          Jackson    Dallas              7500
     4     Christopher   Jones      San Francisco      95000
     5     Mary          Martinez   Los Angeles       150000
```

8.2.5 Mixed Input

You can mix input styles when reading a text file. For example, in CH8_7.TXT, the first field, occupying columns 1 to 14, contains values for the NAME variable. The second field, occupying column 16, contains values for the ETHNIC variable. The last two fields, INCOME (nonstandard numeric value) and CITY (containing embedded blanks), are in free format. Program 8.7 uses the column input method to read the first two fields, NAME and ETHNIC, and then reads INCOME and CITY by using the modified list input method.

CH8_7.txt:

```
123456789012345678901234567890123456789 0
James Miller   W $1,100,000 New York
Patricia Davis B $90,000    Phoenix
John Jackson   W $75,000    Dallas
Mary Martinez  H $150,000   Los Angeles
```

Program 8.7:

```
data ex8_7;
    infile "W:\SAS_Book\dat\ch8_7.txt";
    input name    $ 1 - 14
        ethnic    $ 16
        income    : comma.
        city &    $13.;
run;

title 'Using the mixed input method';
proc print data = ex8_7;
run;
```

Output from Program 8.7:

```
              Using the mixed input method
   Obs          name       ethnic    Income    city
    1       James Miller      W      1100000   New York
    2       Patricia Davis    B        90000   Phoenix
    3       John Jackson      W        75000   Dallas
    4       Mary Martinez     H       150000   Los Angeles
```

8.2.6 Creating Observations by Using the Line Pointer-Controls

In some situations, you may need to create one observation that is based on multiple records in a text file. You can move the input pointer of a specific record by using the following line pointer-controls:

```
/ (forward slash)
#n
```

The forward slash (/) pointer-control moves the input pointer to column 1 of the next record. The #*n* pointer-control moves the input pointer to the record *n*.

Using the forward slash (/) pointer-control enables you to read multiple records sequentially or in any order. To use the line pointer-control, however, you must check that the text file contains the same number of records for each observation. The line pointer-control can be used with any input method.

For example, in the CI I8_8 data set, the information required to create one observation is stored in three records. Program 8.8 uses the forward slash (/) pointer-control to read the data.

CH8_8.txt:

```
123456789012345678901234_5
James Miller
W New York
$1,100,000
Patricia Davis
B Phoenix
$90,000
```

Program 8.8:

```
data ex8_8;
    infile "W:\SAS_Book\dat\ch8_8.txt";
    input fname : $11. lname $/
          ethnic $ city & $/
          income : comma.;
run;

title 'Using forward slash (/) pointer-control';
proc print data = ex8_8;
run;
```

Output from Program 8.8:

```
          Using forward slash (/) pointer-control
     Obs     fname       lname     ethnic      city       income
      1      James       Miller      W        New York   1100000
      2      Patricia    Davis       B        Phoenix      90000
```

Program 8.9 shows how the #*n* pointer-control is applied. First, the second record is read, followed by the first and finally by the last.

Program 8.9:

```
data ex8_9;
    infile "W:\SAS_Book\dat\ch8_8.txt";
    input #2 ethnic $ city & $
          #1 fname : $11. lname $
          #3 income : comma.;
run;

title 'Using #n pointer-control';
proc print data = ex8_9;
run;
```

Output from Program 8.9:

```
                 Using #n pointer-control
     Obs    ethnic    city       fname      lname      income
      1       W      New York    James      Miller    1100000
      2       B      Phoenix     Patricia   Davis       90000
```

8.2.7 Creating Observations by Using Line-Hold Specifiers

The INPUT statement has optional line-hold specifiers that provide greater flexibility for reading a text file into a SAS data set. Line-hold specifiers have either a single trailing "at" sign (@) or double trailing "at" signs (@@). Trailing "at" signs are always placed at the end of an INPUT statement:

```
INPUT <specification(s)><@|@@>;
```

Double trailing "at" signs (@@) are frequently used for creating multiple observations from a single record. With double trailing "at" signs (@@) you can hold data in the input buffer across multiple iterations of the implicit loop of the DATA step. The data record held in the input buffer by double trailing "at" signs (@@) is not released until either the input pointer moves past the end-of-record marker or another INPUT statement without a line-hold specifier executes. You cannot use double trailing "at" signs (@@) with the column input method, @ column pointer-control, or the MISSOVER option in the INFILE statement.

For example, CH8_9.TXT contains two records with each record containing first and last names plus age for two persons. Program 8.10 creates two observations in the SAS data set from each record in the text file by using double trailing "at" signs (@@) in the INPUT statement. During the first iteration of the implicit loop of the DATA step, the INPUT statement copies the first record from CH8_9.TXT to the input buffer. Once the record is copied to the input buffer, the INPUT statement copies the first group of values for FNAME, LNAME, and AGE from the input buffer to the PDV. Then the input pointer is positioned at column 16. The first observation is created at the end of the DATA step. At the beginning of the second iteration of the implicit loop, the double trailing "at" signs (@@) prevent the data value in the input buffer from being released and keep the input pointer at the current location (at column 16). Next, the INPUT statement copies the second group of values in the input buffer to the PDV and the input pointer reaches the *end-of-record* maker. When this happens, the first data record is released from the input buffer. At the beginning of the third iteration of the DATA step, the INPUT statement copies the second record from CH8_9.TXT to the input buffer. The input pointer moves to column 1 of the input buffer and the execution process is repeated until all the values in the second record are read.

CH8_9.txt:

```
123456789012345678901234567890123456789
James Miller 26 Patricia Davis 30
John Jackson 31 Mary Martinez   32
```

Program 8.10:

```
data ex8_10;
    infile "W:\SAS_Book\dat\ch8_9.txt";
    input fname $ lname $ age @@;
run;

title 'Using the double at sign(@@)';
proc print data = ex8_10;
run;
```

Output from Program 8.10:

```
          Using the double at sign(@@)
      Obs    fname       lname      age
      1      James       Miller     26
      2      Patricia    Davis      30
      3      John        Jackson    31
      4      Mary        Martinez   32
```

Similar to the double trailing "at" signs (@@), the single trailing "at" sign (@) enables multiple INPUT statements to be applied to a single record in the input buffer. However, the essential difference between the single and the double trailing "at" signs is that the single trailing "at" sign (@) also releases a record in the input buffer at the beginning of every DATA step execution, while double trailing "at" signs (@@) can hold a record in the input buffer across multiple iterations of the implicit loop of the DATA step until the end-of-record marker is encountered.

   Suppose that you would like to create a SAS data set from CH8_1. TXT but only keep records of the people with AGE > 30. You can use the single trailing "at" sign (@) to accomplish this task. (See Program 8.11.) During the DATA step execution, the first INPUT statement reads the data value from columns 17 and 18 from the input buffer and assigns the AGE variable in the PDV. The single trailing "at" sign (@) holds the records in the input buffer. The subsetting IF statement evaluates the condition (AGE > 30). If the condition is not true, then the current DATA step iteration terminates, and SAS returns to the beginning of the DATA step for the next iteration of the implicit loop while at the same time releasing the contents of the input buffer. In the second iteration, the next record in the input buffer is read in with the first INPUT statement. Again, AGE > 30 is checked, and if TRUE, the second INPUT statement executes with values being copied from the input buffer to the NAME, GENDER, and CITY variables. Because the second INPUT statement doesn't contain a line-hold specifier, the records in the input buffer will then be released after the INPUT statement executes. This process continues until all the records have been processed.

Program 8.11:

```
data ex8_11;
    infile "W:\SAS_Book\dat\ch8_1.txt" truncover;
    input age 17 - 18 @;
    if age > 30;
    input Name $ 1 - 14
          Gender $ 16
          City $ 20 - 30;
run;

title 'Using the single at sign(@)';
proc print data = ex8_11;
run;
```

Output from Program 8.11:

```
                 Using the single at sign(@)
        Obs    age         Name        Gender    City
         1      31     John Jackson       M       Dallas
         2      32     Mary Martinez      F       Los Angeles
```

8.3 Creating Text Files

The idea behind choosing an optimal output method is similar to choosing an input method. For example, to generate a text field in a fixed field, you can use the column output or formatted output methods. Using the formatted output method enables you to associate a format with a variable when you create your output file. To create a free format text file, you can use the list output method. In addition, you can mix different output methods to create a text file. Due to these similarities, generating text files in each of the following sections will not be presented in detail.

8.3.1 Column Output

The column output method is used to write variable values in the specified columns in the output line. The general form of the column output method is as follows:

```
PUT variable start-column <- end-column>;
```

The *variable* is the name of the SAS variable whose value is written to the text file. The *start-column* and the *last-column* are used to specify the first column and the last column of the field where the value is to be written in the text file. The *last-column* can be omitted if the value only copies one column. If the number of columns in a specific value is less than specified, a character variable will be left-aligned at *start-column*, and a numeric variable will be right-aligned at *end-column*. There is no need to write a dollar sign ($) after the character variable in the PUT statement.

In the previous section, Program 8.1 was used to create a SAS data set, EX8_1, with four variables by reading in CH8_1.TXT. The four variables were NAME, GENDER, AGE, and CITY. Program 8.12 shows how to create a text file from EXE8_1 with the column output method that contains fields for only the NAME, AGE, and CITY variables. GENDER is being excluded.

In Program 8.12, a DATA _NULL_ statement is used because the purpose of the DATA step is to create an external text file, not a SAS data set. The SET statement copies the contents from EX8_1 to the PDV. Next, the FILE statement specifies the name of the output file. Then, the PUT statement writes the values from variables NAME, AGE, and CITY to the specified columns in the text file. The generated file, EX8_1_OUT.TXT, is listed after Program 8.12.

Program 8.12:

```
data _null_;
    set ex8_1;
    file "W:\SAS_Book\dat\OUT\ex8_1_out.txt";
    put name 1 - 14
```

```
         age 16 - 17
         city 19 - 29;
run;
```

EX8_1_out.txt:

```
12345678901234567890123456789 0
James Miller    26 New York
Patricia Davis 30 Phoenix
John Jackson    31 Dallas
Mary Martinez   32 Los Angeles
```

8.3.2 Formatted Output

You can use the formatted output method to write formatted variables out to a text file. The general form of the formatted output method is as follows:

```
PUT <pointer-control> variable format;
```

Similar to the formatted input method, two commonly used column pointer-controls are @*n* and +*n*. The @*n* control moves the pointer to column *n*, and +*n* moves the pointer *n* columns forward. You can use the following alignment parameters in the *format* to override the default alignment:

- -L: left-align the value
- -C: center the value
- -R: right-align the value

Program 8.2 creates a SAS data set, EX8_2, by reading in CH8_2.TXT. Program 8.13 re-creates a text file by using the formatted output method. The alignment parameter -L is used to left-align the value of INCOME; otherwise, numeric variables are right-aligned by default.

Program 8.13:

```
data _null_;
    set ex8_2;
    file "W:\SAS_Book\dat\OUT\ex8_2_out.txt";
    put @1 name $14.
        +1 dob mmddyyc8.
        +2 income dollar13.2-l;
run;
```

EX8_2_out.txt:

```
12345678901234567890123456789012345678 90
James Miller   03:02:86 $1,100,000.00
Patricia Davis 11:13:82 $90,000.00
John Jackson   06:15:81 $75,000.00
Mary Martinez  08:02:80 $150,000.00
```

8.3.3 List Output

You can use the list output method to write output to a text file. To use the list output method, simply list the variable(s) after the keyword PUT:

PUT variable;

When the list output method is used, a single-space character is inserted between each variable in the list that follows the PUT statement. Check the output from Program 8.14 for embedded spaces between NAME, AGE, and CITY.

Program 8.14:

```
data _null_;
    set ex8_1;
    file "W:\SAS_Book\dat\OUT\ex8_3_out.txt";
    put name age city;
run;
```

EX8_3_out.txt:

```
12345678901234567890123456789012345678901
James Miller     26 New York
Patricia Davis 30 Phoenix
John Jackson     31 Dallas
Mary Martinez    32 Los Angeles
```

Exercises

Exercise 8.1. Read DAT8_1.TXT into SAS. Descriptions of each field are listed in Table 8.7. The last field of the data contains quotation marks.

TABLE 8.7

Field Description for DAT8_1.TXT

Variable Name	Locations	Variable Type
ID	Columns 1–3	Character
Gender	Column 4	Character
Ethnic	Column 6	Character
City	–	Character
Income	–	Numeric
Smoking	–	Character

Please make sure to remove the quotation marks when you read the data.

DAT8_1.TXT

```
12345678901234567890123456789012345678901234567890123456789
629F H Los Angeles    $35,000.00 "past"
656F W New York       $48,000.00 "never"
711F W Chicago        $30,000.00 "never"
733F W Los Angeles    $59,000.00 "current"
135F B San Francisco $120,000.00 "current"
982F W Boston        $113,000.00 "past"
798F W Chicago        $28,900.00 "never"
494F W Seattle        $65,000.00 "never"
748F W Los Angeles    $39,000.00 "never"
904F W Seattle        $76,200.00 "never"
244F W Orlando        $58,000.00 "never"
747F A New York       $39,000.00 "current"
796F A San Francisco $134,000.00 "past"
713F H Chicago        $29,000.00 "never"
745F A Seattle        $76,000.00 "never"
184M W Boston         $13,900.00 "past"
```

Exercise 8.2. In DAT8_2.TXT, the first two fields contain ID (columns 1–3) and GENDER (column 4), and the last four fields contain quarterly income. Create a SAS data set by reading DAT8_2.TXT. There will be 12 observations in the generated SAS data set and it will contain the following variables: ID, GENDER, QTR (with values 1, 2, 3, or 4), and INCOME (contains monthly income). (Hint: Use a single trailing "at" sign (@) to hold the input pointer.)

DAT8_2.TXT

```
12345678901234567890123456789012345678901234567890123456789
629F $5,000.00 $5,500.00 $4,000.00 $3,500.00
656F $3,400.00 $5,600.00 $7,800.00 $3,100.00
711F $2,800.00 $3,100.00 $2,900.00 $3,800.00
```

The created SAS data set will be similar to the one below:

```
   The SAS data set that created from DAT8_2.TXT
     Obs      id      gender      qrt     income
      1       629        F         1       5000
      2       629        F         2       5500
      3       629        F         3       4000
      4       629        F         4       3500
      5       656        F         1       3400
```

6	656	F	2	5600
7	656	F	3	7800
8	656	F	4	3100
9	711	F	1	2800
10	711	F	2	3100
11	711	F	3	2900
12	711	F	4	3800

9

Data Step Functions

9.1 Introduction to Functions and CALL Routines

SAS® functions and built-in CALL routines perform data manipulation, mathematical calculations, and descriptive statistical computations on argument(s) typically supplied by the user. SAS functions return a single value that can be used in an assignment statement or in additional expressions within the DATA step. In contrast, a CALL routine does not return a value; instead, it only alters the values of its argument. You cannot use a CALL routine in an assignment statement or in an expression.

As of the 9.2 release, SAS provides over 500 functions and CALL routines. Compared to CALL routines, SAS functions are more commonly used in daily programming tasks. Thus, only the introductory material on the CALL routine is presented in this chapter. In addition to SAS documentation, *SAS® Functions by Example* (Cody, 2004) is a good reference for learning SAS functions and CALL routines.

9.1.1 Functions

SAS functions can be used in the DATA step programming statements, WHERE expressions, macro language statements, and the REPORT or SQL procedures. A SAS function typically takes the following form:

```
function-name (argument-1<,...argument-n>)
```

The *argument* in a SAS function can be a variable name, constant, or any valid SAS expression. When a function requires more than one argument, each argument is separated by a comma. Two additional forms are available for programming SAS functions:

```
function-name (OF variable-list)
function-name (<argument | OF variable-list | OF array-name[*]>
       <..., <argument | OF variable-list | OF array-name[*]>>)
```

Based on the syntax above, you can include *variable-list* or *array-name* as the function argument. When using either a variable-list or the name of an array

Output from Program 9.2:

```
Set SCORES to missing if their sum is Greater than 20
    Obs     ID    noise1  noise2  noise3
     1     629F      5       6       9
     2     656F      .       .       .
     3     711F      0       .       3
     4     511F      .       .       .
     5     478F      .       5       3
```

9.1.3 Categories of Functions and CALL Routines

SAS functions and CALL routines can be grouped into more than 20 categories, such as array, descriptive statistics, mathematical, random number, character, date and time, special functions, etc. Three of the array functions (DIM, HBOUND, LBOUND) were presented in Chapter 6. This section covers a few descriptive statistics, mathematical, and random number functions.

SAS provides a large selection of functions for calculating descriptive statistics. The name of the function captures what the function does. For example, the SUM function calculates the sum of its argument or arguments (as the name implies). Some commonly used functions for calculating descriptive statistics are listed in Table 9.1.

Mathematical functions are also available, such as ABS (returns the absolute value), EXP (returns the value of the exponential function),

TABLE 9.1

Descriptive Statistics Functions

Function Name	Returned Value
CMISS	The number of missing arguments
CV	The coefficient of variation
IQR	The interquartile range
LARGEST	The *k*th largest non-missing value
MAX	The largest value
MEAN	The average value
MEDIAN	The median value
MIN	The smallest value
N	The number of non-missing numeric values
NMISS	The number of missing numeric values
PCTL	The percentile that corresponds to the percentage
RANGE	The range of the non-missing values
SUM	The sum of the non-missing arguments
SMALLEST	The *k*th smallest non-missing value
STD	The standard deviation of the non-missing arguments
VAR	The variance of the non-missing arguments

FACT (computes a factorial), GCD (returns the greatest common divisor), LOG (returns the natural logarithm), and SQRT (returns the square root of a value), etc.

The RANUNI function, a random number function, was introduced in Chapter 5. The RANUNI function is used to generate a random number that follows the uniform distribution. In addition, you can generate random numbers from other types of distributions, such as binomial (RANBIN), exponential (RANEXP), normal (RANNOR), Poisson (RANPOI), etc.

9.2 Date and Time Functions

Date and time values are stored as numeric values in SAS. Please refer to Chapter 8 for date and time values, along with date and time formats and informats. Because the dates and times are stored as numeric values internally, you can easily perform computations on date and time values, such as calculating the number of days between two specified dates. However, computations on date and time values can be easily accomplished by using SAS date and time functions.

9.2.1 Creating Date and Time Values

SAS date, time, and datetime values can be created automatically after you read date and time values with appropriate informats. You can create date, time, and datetime constant values directly in the DATA step. The syntax and examples for creating date, time, and datetime constants are listed in Table 9.2.

To obtain the current date, time, and datetime values, you can use the TODAY (or DATE), TIME, and DATETIME functions, respectively, without specifying any arguments. You can also construct a date value by using the MDY and YYQ functions, a time value by using the HMS

TABLE 9.2

Syntaxes and Examples of Date, Time, and Datetime Constants

Constant	Syntax	Example
Date	'ddmmm<yy>yy'D	'18jul2012'd
	"ddmmm<yy>yy"D	"18jul2012"d
Time	'hh:mm<:ss.s>'T	'7:34't
	"hh:mm<:ss.s>"T	"7:34:30am"t
Datetime	'ddmmm<yy>yy:hh:mm<:ss.s>'DT	'18jul2012:7:34'dt
	"ddmmm<yy>yy:hh:mm<:ss.s>"DT	"18jul2012:7:34:30am"dt

TABLE 9.3

Functions for Creating Date and Time Values

Syntax	Value Returned
TODAY()	Current date as a numeric date value
DATE()	Current date as a numeric date value
TIME()	Current time of day as a numeric time value
DATETIME()	Current date and time of day as a datetime value
MDY(*month,day,year*)	Date value from month, day, and year values
YYQ(*year,quarter*)	Date value from year and quarter year values
HMS(*hour,minute,second*)	Time value from hour, minute, and second values
DHMS(*date,hour,minute,second*)	Datetime value from date, hour, minute, and second

function, or the datetime value by using the DHMS function. The syntax of these functions is listed in Table 9.3. Program 9.3 illustrates the use of these functions.

Program 9.3:

```
title 'Constructing date and time values';
data ex9_3;
   current_date = today();
   current_time = time();
   current_datetime = datetime();
   date1 = mdy(8,31,2012);
   date2 = yyq(2012,3);
   time = hms(6,12,30);
   datetime = dhms(current_date, 6,12,30);
   file print;
   put 'current_date:     ' current_date date9.;
   put 'current_time:     ' current_time time.;
   put 'current_datetime: ' current_datetime datetime.;
   put 'date1:            ' date1 date9.;
   put 'date2:            ' date2 date9.;
   put 'time:             ' time time.;
   put 'datetime:         ' datetime datetime.;
run;
```

Output from Program 9.3:

```
                Constructing date and time values
current_date:       18JUL2012
current_time:       9:47:25
current_datetime:   18JUL12:09:47:25
date1:              31AUG2012
date2:              01JUL2012
time:               6:12:30
datetime:           18JUL12:06:12:30
```

9.2.2 Extracting Components from Date and Time Values

In some applications, you might need to know the day, weekday, month, quarter, or year that is based on a single SAS date. These values can be extracted from a date value by using the DAY, WEEKDAY, MONTH, QTR, or YEAR function. Furthermore, you can extract the hour, minute, or second value from either a SAS time or a datetime value. The syntaxes for these functions are listed in Table 9.4.

The next example uses the TENANT data set described in Table 9.5. Program 9.4 shows how to use the YEAR and MONTH functions along with a subsetting IF statement to create a new data set that displays information about those tenants who moved in during August 1987.

TABLE 9.4

Functions for Extracting Components from Data and Time Values

Syntax	Value Returned
YEAR(*date*)	A four-digit numeric value
QTR(*date*)	1–4
MONTH(*date*)	1–12
DAY(*date*)	1–31
WEEKDAY(*date*)	1 (Sunday)–7 (Saturday)
HOUR(*<time | datetime>*)	0–23
MINUTE(*<time | datetime>*)	0–59
SECOND(*<time | datetime>*)	0–<60

TABLE 9.5

Variable Description of TENANT Data

Variable Name	Date Type	Description
ID	Character	Subject ID with four characters. The last character represents gender. For example, 115M
NAME	Character	Subject's full name
AREA_CODE	Character	Three-digit area code
PHONE1	Numeric	The first three digits of the phone number
PHONE2	Numeric	The last four digits of the phone number
MOVE_IN_ DATE	Numeric	A date value representing moving date
CP_MONTH	Numeric	The month (1–12) in which complaints were made about the noise
CP_DAY	Numeric	The day (1–31) on which complaints were made about the noise
COMMENTS	Character	Tenants' comments

Program 9.4:

```
data ex9_4;
    set ch9.tenant;
    if year(move_in_date) = 1987 and month(move_in_date) = 8;
run;

title 'People who moved in Aug 1987';
proc print data = ex9_4;
    var id name move_in_date;
run;
```

Output from Program 9.4:

```
                People who moved in Aug 1987
                                              move_in_
        Obs     id        name                 date
         1     483F    LISA WALKER          10AUG1987
         2     904F    LINDA THOMPSON       05AUG1987
```

9.2.3 Date and Time Interval Functions

You can calculate the number of days in a time interval by subtracting the starting date from the ending date. To obtain the number of years in a time interval, you can divide the number of days in a time interval by 365. Alternatively, you can use the INTCK function to count the number of interval boundaries between two dates, two times, or two datetime values. The basic syntax for the INTCK function is as follows:

INTCK(interval, start-from, increment)

The *interval* argument is used to specify a character constant, a variable, or an expression that contains an interval name. You can use either uppercase or lowercase for the *interval* argument. Examples of the *interval* values are listed in Table 9.6. The *start-from* and *increment* arguments are used to

TABLE 9.6

Examples of Interval Values Used in the INTCK and INTNX Functions

Interval Value	Definition	Default Starting Point
DAY	Daily intervals	Each day
WEEK	Weekly intervals of 7 days	Each Sunday
MONTH	Monthly intervals	First of each month
YEAR	Yearly intervals	January 1
SECOND	Second intervals	Start of the day (midnight)
MINUTE	Minute intervals	Start of the day (midnight)
HOUR	Hourly intervals	Start of the day (midnight)

specify a SAS expression that represents the starting and ending SAS date, time, or datetime values, respectively.

Another useful time-interval function is the INTNX function. The INTNX function is used for generating a date, time, or datetime value by incrementing a date, time, or datetime value by a given time interval. The syntax for the INTNX function is as follows:

```
INTNX(interval, start-from, increment<, 'alignment'>)
```

The use of the *interval* argument is the same as in the INTCK function. The *start-from* argument is used to specify a SAS expression that represents the starting SAS date, time, or datetime value. Unlike the *increment* in the INTCK function, the *increment* in the INTNX function is used to specify a negative, positive, or zero integer that represents the number of intervals to shift the value from the *start-from* argument. The optional *alignment* argument must be enclosed in quotation marks; it is used to control the position of the returned SAS dates within the interval. The *alignment* argument can take one of the following values:

- BEGINNING or B (the default value): The returned date value is aligned to the beginning of the interval.
- MIDDLE or M: The returned date value is aligned to the midpoint of the interval.
- END or E: The returned date value is aligned to the end of the interval.
- SAME or S: The returned date has the same alignment as the input date.

Program 9.5 calculates the number of years since the tenants moved in by using the INTCK function. The INTNX function is used to create a two-year anniversary date from the move-in date.

Program 9.5:

```
data ex9_5;
    set ch9.tenant;
    move_in_years = intck('year', move_in_date, today());
    anniversary2 = intnx('year', move_in_date, 2, 's');
run;

title 'Illustrating the use of INTCK and INTNX functions';
proc print data = ex9_5 (obs = 5);
    var move_in_date move_in_years anniversary2;
    format anniversary2 date9.;
run;
```

Output from Program 9.5:

```
     Illustrating the use of INTCK and INTNX functions
           move_in_  move_in_
    Obs       date     years  anniversary2
     1      09SEP1989    23     09SEP1991
     2      05MAY1995    17     05MAY1997
     3      08NOV1995    17     08NOV1997
     4      03FEB1991    21     03FEB1993
     5      15FEB2000    12     15FEB2002
```

9.3 Character Functions

Often you need to change the case of a text string in SAS. You also might want to concatenate two strings or extract a substring of characters from a larger character string. All these tasks, plus character string searches, can easily be handled by character functions in SAS. Many are available. A few are described below.

9.3.1 Functions for Changing Character Cases

The UPCASE and the LOWCASE functions can change all letters in its argument to uppercase and lowercase, respectively. The syntax for these two functions is as follows:

UPCASE (argument)
LOWCASE (argument)

The *argument* in these two functions is used to specify a character constant, variable, or expression. If the UPCASE or LOWCASE functions return a value to a variable that has not been previously assigned with a length, the variable will be given the length of the *argument*.

The PROPCASE function can be used to change the words in its argument to proper case. The syntax for the PROPCASE function is as follows:

PROPCASE (argument <,delimiters>)

The PROPCASE function first converts all uppercase letters in the argument to lowercase and then converts the first character of a *word* to uppercase. A word is defined as a character string preceded by the value specified in the *delimiters* argument. The default value for the delimiter is a blank, forward slash, hyphen, open parenthesis, period, or tab. If the PROPCASE function returns a value to a variable that has not been previously assigned a length, that variable is given a length of 200 bytes.

In Program 9.6, the NEW_NAME variable is created by changing the NAME variable to its proper case by using the PROPCASE function. NEW_COMMENTS is generated by changing each value for COMMENTS to uppercase with the UPCASE function.

Program 9.6:

```
data ex9_6;
    set ch9.tenant;
    new_name = propcase(name);
    new_comments = upcase(comments);
run;

title 'Illustrating the use of PROPCASE and UPCASE functions';
proc print data = ex9_6 (obs = 5);
    var name new_name comments new_comments;
run;
```

Output from Program 9.6:

```
        Illustrating the use of PROPCASE and UPCASE functions
Obs       name            new_name               comments
 1    DAVID DAVIS     David Davis      pounding; complains often
 2    DANIEL THOMAS   Daniel Thomas    sleep OK
 3    JESSICA SCOTT   Jessica Scott    Booming; complains often
 4    THOMAS TAYLOR   Thomas Taylor    Pounding
 5    MARGARET LEWIS  Margaret         Lewis Boom


Obs           new_comments
 1      POUNDING; COMPLAINS OFTEN
 2      SLEEP OK
 3      BOOMING; COMPLAINS OFTEN
 4      POUNDING
 5      BOOM
```

9.3.2 Functions for Concatenating Character Strings

One way to concatenate characters is to use the concatenation operator (||). The concatenation operator is positioned between the two character variables being concatenated. You can use an assignment statement to store the concatenated strings into a new variable. The length of the resulting variable is the sum of the lengths of each variable or constant in the concatenation operation. Note that the concatenation operator does not trim the leading and trailing blanks in the variables or constants being concatenated. Program 9.7 uses the concatenation operator to concatenate the variables AREA_CODE, PHONE1, PHONE2, and dash characters ('-'). Results are stored in the PHONE variable.

Program 9.7:

```
data ex9_7;
    set ch9.tenant;
    phone = area_code||'-'||phone1||'-'||phone2;
run;

title 'Illustrating the use of the concatenating operator';
proc print data = ex9_7(obs = 5);
    var area_code phone1 phone2 phone;
run;
```

Output from Program 9.7:

```
       Illustrating the use of the concatenating operator
       area_
Obs    code   phone1    phone2          phone
 1     714     545       3799      714-   545-   3799
 2     714     179       5944      714-   179-   5944
 3     714     640       6869      714-   640-   6869
 4     310     165       7540      310-   165-   7540
 5     714     426       2065      714-   426-   2065
```

The variable AREA_CODE is a character variable, but PHONE1 and PHONE2 are numeric variables in the TENANT data set. Because the concatenation operator requires the variables on both sides of the concatenation operator to be characters, SAS performs automatic numeric-to-character conversion by using the BEST12. format. The resulting character values will have leading blanks after the conversion because BEST12. supports right justification. The large gaps within the newly created PHONE variable can be removed by using one or a combination of character alignment functions described in Table 9.7.

Instead of using the concatenation operator, you can use the more convenient CAT, CATT, CATS, and CATX functions to concatenate character strings. The CAT function does not remove leading and trailing blanks in the resulting concatenated character string. The CATT function removes only the trailing blanks, and the CATS function removes both leading and

TABLE 9.7

Functions for Aligning Character Strings and Trimming Blanks

Syntax	Description
LEFT(*argument*)	Left-aligns a character string
RIGHT(*argument*)	Right-aligns a character string
TRIM(*argument*)	Removes trailing blanks from a character string and returns one blank if a string is missing
STRIP(*argument*)	Returns a character string with all leading and trailing blanks removed

TABLE 9.8

Functions for Aligning Character Strings and Trimming Blanks

Example	Equivalent Code
`CAT(OF V1-V3)`	`V1||V2||V3`
`CATT(OF V1-V3)`	`TRIM(V1)||TRIM(V2)||TRIM(V3)`
`CATS(OF V1-V3)`	`TRIM(LEFT(V1))||TRIM((LEFT(V2))||` `TRIM(Left(V3))`
`CATX('-',OF V1-V3)`	`TRIM(LEFT(V1))||'-'||TRIM((LEFT(V2))||` `'-'||TRIM(Left(V3))`

trailing blanks. The CATX function not only removes both the leading and trailing blanks, it also inserts a *delimiter* between each concatenating item. The syntax for these functions is listed below; notice that they are almost identical. The *item*(s) are used to specify a constant, a variable, or an expression; they can be either character or numeric. If *item* is numeric, it will be converted to a character by using the BEST*w*. format. Some examples for using these functions are listed in Table 9.8.

```
CAT(item-1 <,..., item-n>)
CATT(item-1 <,..., item-n>)
CATS(item-1 <,..., item-n>)
CATX(delimiter, item-1 <,...item-n>)
```

Program 9.8 illustrates how to eliminate the gaps in the concatenated result. The first method uses both the LEFT and TRIM functions along with the concatenation operator. The second method uses the STRIP function with the concatenation operator. The last method uses the CATX function only. All these methods return the same result.

Program 9.8:

```
data ex9_8;
    set ch9.tenant;
    phone_number1 = area_code||'-'||trim(left(phone1))
                    ||'-'||trim(left(phone2));
    phone_number2 = area_code||'-'||strip(phone1)||'-'||
                    strip(phone2);
    phone_number3 = catx('-', area_code, phone1, phone2);
run;

title 'Eliminating the gaps in the PHONE variable';
proc print data = ex9_8 (obs = 5);
    var area_code phone1 phone2 phone_number1
        phone_number2 phone_number3;
run;
```

Output from Program 9.8:

```
         Eliminating the gaps in the PHONE variable
        area_                   phone_        phone_        phone_
Obs  code phone1 phone2    number1       number2       number3
 1   714    545    3799  714-545-3799  714-545-3799 714-545-3799
 2   714    179    5944  714-179-5944  714-179-5944 714-179-5944
 3   714    640    6869  714-640-6869  714-640-6869 714-640-6869
 4   310    165    7540  310-165-7540  310-165-7540 310-165-7540
 5   714    426    2065  714-426-2065  714-426-2065 714-426-2065
```

9.3.3 Functions for Searching, Exacting, and Replacing Character Strings

There is a large selection of functions that can be used for searching, extracting, and replacing character strings. For example, to search a character expression for a string of characters, you can use the INDEX function, which has the following form:

INDEX(source, excerpt)

Both the *source* and *excerpt* arguments in the INDEX function are used to specify a character constant, variable, or expression. The INDEX function searches *source* from left to right for the first occurrence of the character string that is specified in the *excerpt* argument. If the character string is found, the INDEX function will return the position in *source* of the string's first character. If the character string is not found, the INDEX function will return 0. Note that if there are multiple occurrences of the string, the INDEX function returns only the position of the first occurrence. When a variable is used in the *excerpt* argument, both leading and trailing spaces are also considered part of the *excerpt* argument. You can remove the spaces by using the TRIM or STRIP function with the *excerpt* argument within the INDEX function.

To extract a substring from a character string, you can use either the SCAN or SUBSTR(right of =) functions. The SCAN function is used to extract the *n*th word from a character string. A word is defined as a substring that is separated by delimiters. The default length of the variable that stores the results from a SCAN is 200 bytes unless the variable has been previously defined by the LENGTH statement. The basic form of the SCAN function is as follows:

SCAN(string, count <,charlist >)

The *string* argument is used to specify a character constant, variable, or expression. The *count* argument is a nonzero integer, variable, or expression that has a nonzero integer value. You use *count* to identify the word by number in *string* that you want to extract. For example, when *count* is set to 1, the first word is returned, 2 returns the second word, and so on. If *count* is positive, the SCAN function counts words from left to right in *string*, and if *count* is negative, the SCAN function will count from right to left. The optional *charlist* argument in the SCAN function is used to identify delimiters that separate *string* into words. If the *charlist* is not specified, the default delimiter (in ASCII environments) is used, including blank ! $% & () * +, -./; < ^ |.

Unlike the SCAN function, the SUBSTR(right of =) function requires that you supply a starting position plus a number of characters to extract a designated substring. The syntax is as follows:

```
<variable=>SUBSTR(string, position<,length>)
```

The reason for listing the *variable* = on the left side of the SUBSTR(right of =) function is to distinguish the SUBSTR(right of =) function from the SUBSTR(left of =) function that is presented next. The *string* argument is used to specify a character constant, variable, or expression from which you want to extract a substring. The *position* and the *length* argument can be a numeric constant, variable, or expression. The *position* is the starting position that you want to extract, and the *length* argument is the length of the substring to extract. If *length* is omitted, SAS will extract the remainder of *string*. The length of the target *variable* will be the length of *string* unless a length has been previously assigned to *variable*.

Program 9.9 illustrates the use of the SCAN, SUBSTR(right of =), and INDEX functions. The SCAN function is used to create the FNAME (first name) and LNAME (last name) variables by extracting the first and second words from the NAME variable. Because the first and last names are separated by blanks, which is a default delimiter, the *charlist* argument is not specified in the SCAN function. Next, the NEWID and GENDER variables are created by using the SUBSTR(right of =) function. NEWID is created by subtracting the first three characters from the ID variable. GENDER is created by subtracting the last character. The INDEX function is used with the subsetting IF statement to keep only the observations with COMMENTS variables that contain 'sleep ok' character strings. The LOWCASE function nested within the INDEX function ensures 'sleep ok' in COMMENTS will retain its original case setting.

Program 9.9:

```
data ex9_9;
    set ch9.tenant;
    length fname lname $ 20 newid $ 3 gender $ 1;
    fname = scan(name, 1);
    lname = scan(name, 2);
    newid = substr(id, 1, 3);
    gender = substr(id, 4);
    if index(lowcase(comments),'sleep ok') > 0;
run;
```

```
title 'The use of SCAN, SUBSTR(right of =) and INDEX functions';
proc print data = ex9_9;
    var name fname lname id newid gender comments;
run;
```

Output from Program 9.9:

```
The use of SCAN, SUBSTR(right of =) and INDEX functions
      Obs          name            fname       lname
       1      DANIEL  THOMAS      DANIEL       THOMAS
       2      WILLIAM BROWN       WILLIAM      BROWN
       3      MICHAEL JONES       MICHAEL      JONES
       4      CHARLES WILSON      CHARLES      WILSON
       5      BETTY YOUNG         BETTY        YOUNG
       6      DOROTHY LEE         DOROTHY      LEE
       7      DEBORAH HILL        DEBORAH      HILL
       8      MARK WHITE          MARK         WHITE
       9      MARIA CLARK         MARIA        CLARK
      10      LINDA MILLER        LINDA        MILLER
      11      SANDRA KING         SANDRA       KING

      Obs     id      newid    gender    comments
       1     135M      135       M        sleep OK
       2     478M      478       M        Booming; sleep OK
       3     511M      511       M        sleep OK
       4     590M      590       M        sleep OK
       5     713F      713       F        sleep OK
       6     792F      792       F        sleep OK
       7     793F      793       F        sleep OK
       8     798M      798       M        Booming; sleep OK
       9     823F      823       F        sleep OK
      10     904F      904       F        sleep OK
      11     927F      927       F        Booming; sleep OK
```

In some applications, you may need to replace specific characters or a substring within a string of characters with some character values. For example, you can use the SUBSTR (left of =) function to replace character value contents, which has the following form:

SUBSTR(variable, position<,length>) = characters-to-replace

Using the SUBSTR(left of =) function is similar to the SUBSTR(right of =) function except that SUBSTR(left of =) is placed on the left side of the equal sign. The *character-to-replace*, which is on the right side of the equal sign, can be a character constant, variable, or expression for replacing the contents of *variable*.

Another useful character replacement function is TRANWRD, which replaces all occurrences of a substring in a character string. The syntax for the TRANWRD function is as follows:

TRANWRD(source, target, replacement)

All three arguments in the TRANWRD function can be a character constant, variable, or expression. The *source* argument is the string that you want to translate. The *target* argument, with length greater than 0, is a substring that you search for in the *source* argument. The *replacement* is used to replace the *target* in the *source* argument. If the *replacement* string has zero length, the TRANWRD function will use a single blank to replace the *target*. By default, if the TRANWRD function returns a value and is assigned to a variable, this variable will have a default length of 200 bytes.

In the TENANT data set, the last character of the ID is either 'M' or 'F'. Suppose that you would like to replace 'M' with '1' and 'F' with '0'. You can first use the SUBSTR(right of =) function to compare the last character in the ID variable to see if it equals 'M'. If the last character of the ID equals 'M', then you can use the SUBSTR(left of =) function to replace the last character in the ID variable with '1'; otherwise, replace the last character with '0'. For comparison purposes, the ID2 variable is created by copying the values from ID in Program 9.10 first, and then the last character in ID2 is replaced with either '1' or '0'. Program 9.10 also creates a variable, NEW_COMMENTS, by replacing the string 'Booming' with 'Boom'.

Program 9.10:

```
data ex9_10;
    set ch9.tenant;
    id2 = id;
    if substr(id2, 4) = "M" then substr(id2, 4) = "1";
    else substr(id2, 4) = "0";
    new_comments = tranwrd(comments, 'Booming', 'Boom');
run;

title 'The use of SUBSTR(left of =) and TRANWRD functions';
proc print data = ex9_10 (obs = 5);
    var id id2 comments new_comments;
run;
```

Output from Program 9.10:

```
        The use of SUBSTR(left of =) and TRANWRD functions
Obs id    id2      comments                  new_comments
1   115M  1151  pounding; complains often  pounding; complains
                                           often
2   135M  1351  sleep OK                   sleep OK
3   184F  1840  Booming; complains often   Boom; complains
                                           often
4   188M  1881  Pounding                   Pounding
5   244F  2440  Boom                       Boom
```

9.4 Functions for Converting Variable Types

Sometimes a variable that produces what appears to be numeric output is actually typed as a character. In this situation, you may want to use the INPUT *function* to convert a string into a number. On other occasions you may want to reverse the direction of your conversion so that a number is recast as a character string. PUT, the inverse of INPUT, is what is needed in this instance to do the conversion.

9.4.1 The INPUT Function

In Program 9.11, both INCOME1 and INCOME2 are entered as character variables. In the second DATA step, MONTH_INCOME1 and MONTH_INCOME2 are calculated from INCOME1 and INCOME2. Based on the output generated from Program 9.11, values for MONTH_INCOME1 are correct, whereas MONTH_INCOME2 contains only missing values.

Program 9.11:

```
data income;
    input name $ income1 $ income2 $;
datalines;
John 123000 123,000
Mary 131000 131,000
;
data ex9_11;
    set income;
    month_income1 = income1/12;
    month_income2 = income2/12;
run;

title 'Automatic character-to-numeric conversion';
proc print data = ex9_11;
run;
```

Output from Program 9.11:

```
        Automatic character-to-numeric conversion
                                    month_     month_
   Obs  name    income1   income2   income1    income2
    1   John    123000    123,000   10250.00      .
    2   Mary    131000    131,000   10916.67      .
```

Partial Log from Program 9.11:

```
2515 data ex9_11;
2516    set income;
```

```
2517    month_income1 = income1/12;
2518    month_income2 = income2/12;
2519 run;
```

```
NOTE: Character values have been converted to numeric
      values at the places given by: (Line):(Column).
      2517:21  2518:21
NOTE: Invalid numeric data, income2 = '123,000', at line 2518
      column 21.
name = John income1 = 123000 income2 = 123,000
      month_income1 = 10250
month_income2 =.  _ERROR_ = 1 _N_ = 1
NOTE: Invalid numeric data, income2 = '131,000', at line 2518
      column 21.
name = Mary income1 = 131000 income2 = 131,000
month_income1 = 10916.666667 month_income2 =. _ERROR_ = 1 _N_ = 2
NOTE: Missing values were generated as a result of performing
      an operation on missing values.
      Each place is given by:
      (Number of times) at (Line):(Column).
      2 at 2518:28
NOTE: There were 2 observations read from the data set
      WORK.INCOME.
NOTE: The data set WORK.EX9_11 has 2 observations and 5
      variables.
NOTE: DATA statement used (Total process time):
      real time        0.00 seconds
      cpu time         0.01 seconds
```

When the second DATA step in Program 9.11 executes, SAS automatically converts the character value in INCOME1 to a numeric value first before dividing it by 12. When the automatic conversion occurs, SAS will issue a message in the log indicating that the conversion has occurred. (See the partial log from Program 9.11.) However, automatic conversion doesn't work for INCOME2.

Automatic conversion occurs only when a character value is assigned to a previously defined numeric variable. Automatic conversion also occurs when a character value is used in an arithmetic operation, compared to a numeric value by using a comparison operator, or specified in a function that requires a numeric argument.

Automatic conversion uses the *w.d* informat. Therefore, if a character value does not conform to standard numeric notation, like the values for INCOME2 with their embedded commas, the conversion will yield to a missing value. In this instance, use the INPUT function to perform the proper character-to-numeric conversion. The INPUT *function* has the following form:

INPUT(source, informat)

The *source* argument is used to specify a character constant, variable, or expression to which you want to apply an *informat*. The INPUT statement requires that *source* be typed as a character. The *informat* argument is a SAS *informat* that is applied to *source*. Program 9.12 uses the INPUT function to convert both INCOME1 and INCOME2 from character to numeric before any calculations are made. This time, the correct output is generated.

Program 9.12:

```
data ex9_12;
    set income;
    month_income1 = input(income1, 6.)/12;
    month_income2 = input(income2, comma7.)/12;
run;

title 'Using INPUT function to convert character values to
numeric';
proc print data = ex9_12;
run;
```

Output from Program 9.12:

```
Using INPUT function to convert character values to numeric
                                        month_        month_
    Obs    name    income1    income2    income1       income2
     1     John     123000    123,000   10250.00      10250.00
     2     Mary     131000    131,000   10916.67      10916.67
```

When using the INPUT function, you can specify either a numeric or character *informat*. If a character informat is assigned to *informat*, character output is generated, meaning that a character-to-character conversion takes place. For example, in Program 9.13, the INPUT function is used with the *$UPCASEw.* informat to convert mixed-case character strings to uppercase. Note that a character *informat* is referenced because the argument has a leading dollar sign ($).

Program 9.13:

```
data ex9_13;
    set ch9.tenant;
    new_comments = input(comments, $upcase30.);
run;

title 'Using INPUT function to convert character values to
uppercase';
proc print data = ex9_13 (obs = 5);
    var comments new_comments;
run;
```

Output from Program 9.13:

```
Using INPUT function to convert character values to uppercase
Obs   comments                           new_comments
  1 pounding; complains often            POUNDING; COMPLAINS OFTEN
  2 sleep OK                             SLEEP OK
  3 Booming; complains often BOOMING; COMPLAINS OFTEN
  4 Pounding                             POUNDING
  5 Boom                                 BOOM
```

9.4.2 The PUT Function

SAS can also perform automatic numeric-to-character conversion as illustrated in Program 9.7 where values for PHONE1 and PHONE2 are converted to character strings during the concatenation operation. Similar to the automatic character-to-numeric conversion, SAS will also issue a message in the log when automatic numeric-to-character conversion occurs.

Automatic numeric-to-character conversion occurs when a numeric value is assigned to a previously defined character variable. Automatic conversion also occurs when an *operator* requires a character value (like the concatenation operator '||'), or when a *function* requires a character argument. Instead of relying upon an automatic numeric-to-character conversion, use the PUT function to convert numeric values to character strings. The PUT function has the following form:

PUT(source, format)

The *source* argument is a constant, variable, or expression whose value you want to reformat. The *source* argument can be either character or numeric. The *format* argument is used to specify the SAS format that you want applied to the value in the *source* argument. *Format* must have the same type as the *source*. For example, if the *source* is character, you will need to use a character format denoted by a leading dollar sign ($); on the other hand, if the *source* is numeric, *format* also has to be numeric. Output from a PUT statement is always character, however, because character and numeric formats always generate character output. Program 9.14 uses the PUT function to apply a numeric-to-character conversion on PHONE1 and PHONE2 before the concatenation operator is applied.

Program 9.14:

```
data ex9_14;
    set ch9.tenant;
    phone_number = area_code||'-'||put(phone1, 3.)||'-'
                   ||put(phone2, 4.);
run;
```

```
title 'Using PUT function to convert numeric values to
character';
proc print data = ex9_14 (obs = 5);
    var area_code phone1 phone2 phone_number;
run;
```

Output from Program 9.14:

```
Using PUT function to convert numeric values to character
            area_
    Obs     code    phone1    phone2     phone_number
     1       714      545      3799      714-545-3799
     2       714      179      5944      714-179-5944
     3       714      640      6869      714-640-6869
     4       310      165      7540      310-165-7540
     5       714      426      2065      714-426-2065
```

In Program 9.13, the INPUT function is used with the *$UPCASEw. informat* to create a new variable, NEW_COMMENTS. You can also use the PUT function to achieve the same result with the *$UPCASEw.* format. (See Program 9.15.)

Program 9.15:

```
data ex9_15;
    set ch9.tenant;
    new_comments = put(comments, $upcase30.);
run;


title 'Using PUT function to reformat the COMMENTS variable';
proc print data = ex9_15 (obs = 5);
    var comments new_comments;
run;
```

Output from Program 9.15:

```
        Using PUT function to reformat the COMMENTS variable
Obs  comments                           new_comments
 1  pounding; complains often           POUNDING; COMPLAINS OFTEN
 2  sleep OK                            SLEEP OK
 3  Booming; complains often BOOMING; COMPLAINS OFTEN
 4  Pounding                            POUNDING
 5  Boom                               BOOM
```

Knowing when to use INPUT and PUT functions can be confusing. Table 9.9 summarizes the data types used in the *source* arguments and *returned* values for the INPUT and PUT functions. The *source* for INPUT must be in character form, whereas the *source* for PUT can be either numeric or character. On the other hand, *returned values* from INPUT can be either numeric or character, whereas PUT always returns a character value.

TABLE 9.9

Type of *Source, Informat/Format,* and Returned Values in
the INPUT/PUT Functions

Function	Source	Informat/Format	Returned Value
INPUT(*source, informat*)	Character	Numeric informat	Numeric
	Character	Character informat	Character
PUT(*source, format*)	Numeric	Numeric format	Character
	Character	Character format	Character

TABLE 9.10

Field Description for the ID Variable

Field	Contents
First field	CONSENT: 1 or 0
Second field	GENDER: M or F
Varies: Immediate after the second ID field with length equals 3 to 6	ID
Last field	COUNTY: LA, SB, or OC (Case Varies) Some observations don't have county information

Exercises

Exercise 9.1. In CH9_Q1.SAS7BDAT, there is only one variable named ID. Each field of ID is described in Table 9.10. For this exercise, create the following variables:

- CONSENT: With values equaling 1 or 0.
- GENDER: With values equaling M or F.
- PATIENTID: With length equaling 6. Because the Patient ID with length varies from 3 to 6 from the original ID variable, add zeros at the very beginning. For example, if the original Patient ID is 123, then the newly created variable will be 000123; if the original Patient ID is 1234, then the new value will be 001234; etc.
- COUNTY: With values equaling LA, SB, or OC.

Exercise 9.2. The LAG function is used to retrieve values from previous observations, whereas DIF calculates the difference between the value from the current and previous observations. Often these two special functions are used together to simplify DATA step programming. In this exercise, read the SAS documentation on the LAG and DIFF functions and then create two variables,

HEIGHT_LAST and HEIGHT_DIFF, by processing the HEIGHT. SAS7BDAT data set. HEIGHT_LAST will contain the heights of students in their previous grade, and HEIGHT_DIFF will contain the differences in height between their current and previous grade. The new output data set will look like the one below:

	NAME	GRADE	HEIGHT	HEIGHT_LAST	HEIGHT_DIFF
1	John	6	58.5	.	.
2	John	7	58.8	58.5	0.3
3	John	8	59.2	58.8	0.4
4	Mary	6	57.3	.	.
5	Mary	7	58.0	57.3	0.7
6	Mary	8	60.1	58.0	2.1

10

Useful SAS® Procedures

10.1 Using the SORT Procedure to Eliminate Duplicate Observations

A data set often contains duplicate values across observations. Individual records can be duplicated or a subset of variables among records can have identical values. One way to eliminate duplicate records from a data set is to use FIRST.VARIABLE and LAST.VARIABLE within the DATA step, which was illustrated in Section 4.2.2. If you don't want to alter the composition of your original data set, however, you can use the NODUPKEY or the NODUPRECS options in the SORT procedure and send the output to a second data set that is referenced in the OUT = option. The form for PROC SORT with these options is as follows:

```
PROC SORT  <DATA=SAS-data-set>
           <OUT=SAS-data-set>
           <NODUPKEY>
           <NODUPRECS>;
   BY <DESCENDING> variable-1 <... <DESCENDING> variable-n>;
RUN;
```

10.1.1 Eliminating Observations with Duplicate BY Values

Consider the data set DUPLICATES, which contains three variables: ID, GRADE, and SCORE. Depending upon the BY variable being examined, the number of observations with the same BY values varies. For example, if ID is the BY variable, there are four observations (observations 1, 2, 4, and 6) with identical 'A01' values and three with identical 'A02' values.

DUPLICATES:

	ID	GRADE	SCORE
1	A01	A	3
2	A01	B	3
3	A02	A	4

(Continued)

205

	ID	GRADE	SCORE
4	A01	A	3
5	A02	B	2
6	A01	B	2
7	A02	B	2

When the NODUPKEY option in PROC SORT is used, observations with duplicate BY values are eliminated. PROC SORT sorts first and then compares BY values for a given observation to what is recorded in the previous observation. When a match is found, the current observation is not written to the output data set. Program 10.1 illustrates the different results by using different combinations of BY variables in PROC SORT. In the last PROC SORT, the keyword _ALL_ is used to sort all the variables in the input data set.

Program 10.1:

```
libname ch10 'W:\SAS Book\dat';
proc sort data = ch10.duplicates out = by_id_1 nodupkey;
    by id;
run;

proc sort data = ch10.duplicates out = by_id_grade_1 nodupkey;
    by id Grade;
run;

proc sort data = ch10.duplicates out = by_all_1 nodupkey;
    by _all_;
run;

title 'NODUPKEY: by ID';
proc print data = by_id_1;
run;

title 'NODUPKEY: by ID and GRADE';
proc print data = by_id_grade_1;
run;

title 'NODUPKEY: by ID, GRADE, and SCORE';
proc print data = by_all_1;
run;
```

Output from Program 10.1:

```
              NODUPKEY: by ID
       Obs      ID       grade      score
        1       A01        A          3
        2       A02        A          4
```

```
            NODUPKEY: by ID and GRADE
     Obs        ID        grade      score
      1        A01          A          3
      2        A01          B          3
      3        A02          A          4
      4        A02          B          2


         NODUPKEY: by ID, GRADE, and SCORE
      Obs        ID        grade      score
       1        A01          A          3
       2        A01          B          2
       3        A01          B          3
       4        A02          A          4
       5        A02          B          2
```

10.1.2 Eliminating Duplicate Observations

When using the NODUPRECS or NODUP options, PROC SORT eliminates duplicate observations. PROC SORT compares all variable values for each observation to the ones from the previous observation. If a match is found, then the observation is not written to the output data set.

PROC SORT compares only adjacent observations for eliminating observations. To correctly eliminate all the duplicate observations, all the variables in the input set need to be sorted. In the DUPLICATES data set, there are two identical observations for ID equals 'A01' and two identical observations for 'A02'. Program 10.2 illustrates what happens when different variables are used in the BY statement with the NODUP option. In the first PROC SORT, a duplicate observation is not eliminated because two identical observations for 'A01' do not follow each other after sorting ID. In the second PROC SORT, all duplicate records are eliminated because BY now sorts by ID and within each ID by GRADE. Similarly, sorting all the variables in the last PROC SORT will also eliminate all duplicates. Thus, to guarantee the elimination of duplicate observations from a data set, sort BY _ALL_ and use either the NODUPRECS or NODUP option.

Program 10.2:

```
proc sort data = ch10.duplicates out = by_id_2 nodup;
    by id;
run;

proc sort data = ch10.duplicates out = by_id_grade_2 nodup;
    by id Grade;
run;

proc sort data = ch10.duplicates out = by_all_2 nodup;
    by _all_;
run;
```

```
title 'NODUP: by ID';
proc print data = by_id_2;
run;

title 'NODUP: by ID and GRADE';
proc print data = by_id_grade_2;
run;

title 'NODUP: by ID, GRADE, and SCORE';
proc print data = by_all_2;
run;
```

Output from Program 10.2:

```
                  NODUP: by ID
         Obs      ID      grade    score
          1       A01       A        3
          2       A01       B        3
          3       A01       A        3
          4       A01       B        2
          5       A02       A        4
          6       A02       B        2

             NODUP: by ID and GRADE
         Obs      ID      grade    score
          1       A01       A        3
          2       A01       B        3
          3       A01       B        2
          4       A02       A        4
          5       A02       B        2

        NODUP: by ID, GRADE, and SCORE
         Obs      ID      grade    score
          1       A01       A        3
          2       A01       B        2
          3       A01       B        3
          4       A02       A        4
          5       A02       B        2
```

10.2 Using the COMPARE Procedure to Compare the Contents of Two Data Sets

Sometimes two data sets need to be compared to see if they are the same or if they are different. Instead of using DATA step programming, you can use the COMPARE procedure to compare the contents of two data sets or selected variables from different data sets or within the same data sets. The syntax for PROC COMPARE is listed below:

```
PROC COMPARE <BASE=SAS-data-set>
             <COMPARE=SAS-data-set>
             <BRIEFSUMMARY>;
   BY <DESCENDING> variable-1 <...<DESCENDING> variable-n>;
   ID <DESCENDING> variable-1 <...<DESCENDING> variable-n>;
   VAR variable(s);
   WITH variable(s);
RUN;
```

There are a myriad of options available in PROC COMPARE for controlling how values are compared and how reports are generated. Only a few options are introduced in this chapter. Readers should refer to SAS documentation for additional information.

10.2.1 Information Provided from PROC COMPARE

PROC COMPARE compares the *base* data set with the *comparison* data set specified in the BASE= and COMPARE= options, respectively. PROC COMPARE starts with comparing data set attributes, examining whether both data sets contain matching variables (variables with the same variable name), checking attributes of the matching variables, and determining whether both data sets have matching observations. The matching variables must have the same data type. The matching observations either have the same values for variables specified in the ID statement or occur in the same position in the data sets. After making these comparisons, PROC COMPARE compares the values in the portions of the data sets that match.

In Program 10.3, two data sets are created: COMPARE1 and COMPARE2. Four variables in both COMPARE1 and COMPARE2 have the same variable names: STUDENTID, GENDER, HEIGHT, and WEIGHT1. Among these four variables, HEIGHT has character data type in the COMPARE1 data set, but HEIGHT has numeric data type in the COMPARE2 data set. Another difference between these two data sets is that the WEIGHT2 variable exists only in the COMPARE1 data set. Also, variable WEIGHT1 is assigned with numeric 6.2 format and labeled with "first weight value" only in COMPARE1. The last difference between these two data sets is that there are four observations in COMPARE1 but only three observations in COMPARE2.

Program 10.3 uses only the BASE= and COMPARE= options to compare the COMPARE1 and COMPARE2 data sets without utilizing other options. Because the ID statement is not used in the program, values in COMPARE1 and COMPARE2 are compared based on matching variables with the same name and type, and the matching observations occur in the same position of the data set. That is to say, only values in the first three observations in COMPARE1 and all the observations in COMPARE2 in STUDENTID, GENDER, and WEIGHT1 variables are compared.

In the generated output from PROC COMPARE, the "Variable Summary" table summarizes variable differences, including the number of common variables (4), the number of variables in one data set but not in the other (one in COMPARE1 but not in COMPARE2), the number of variables that have conflicting types (1), and the number of variables that have different attributes (1). The variable with conflicting type (HEIGHT) appears in the "Listing of Common Variables with Conflicting Types" table. The variable with different attributes (WEIGHT1) appears in the "Listing of Common Variables with Differing Attributes" table.

The "Observation Summary" table summarizes observation differences, such as the number of observations in common (3), the number of observations in one data set but not in the other (1), the number of observations that have unequal values in some compared variables (2), the number of observations that have equal values for all compared variables (1), etc.

The "Value Comparison Summary" table lists the number of variables that have equal values across all comparable observations (1), the number of variables that have unequal values with some observations (2), the total number of unequal values (2), and the maximum difference in unequal numerical values (0.1).

The variables with unequal values (STUDENTID and WEIGHT1) are listed in the "Variables with Unequal Values" table. All the values that have differences in the two data sets are listed in the "Value Comparison Results for Variables" table.

Program 10.3:

```
data compare1;
    input studentID $ gender $ height $ weight1 weight2;
    format weight1 6.2;
    label weight1 = first weight value;
    datalines;
A01 M 67 154.3 154.7
A02 M 64 150.2 150.2
A03 F 58 149.1 149.2
A04 F 59 149.2 149.2
;

data compare2;
    input studentID $ gender $ height weight1;
    datalines;
A01 M 67 154.3
A02 M 64 150.3
A04 F 59 149.1
;
```

```
title 'PROC COMPARE: basic output';
proc compare base = compare1
            compare = compare2;
run;
```

Output from Program 10.3:

```
                    PROC COMPARE: basic output
                     The COMPARE Procedure
           Comparison of WORK.COMPARE1 with WORK.COMPARE2
                        (Method = EXACT)
                       Data Set Summary
Dataset                  Created          Modified   NVar  NObs
WORK.COMPARE1   06AUG12:16:41:03   06AUG12:16:41:03    5    4
WORK.COMPARE2   06AUG12:16:41:03   06AUG12:16:41:03    4    3

                       Variables Summary
Number of Variables in Common: 4.
Number of Variables in WORK.COMPARE1 but not in WORK.
   COMPARE2: 1.
Number of Variables with Conflicting Types: 1.
Number of Variables with Differing Attributes: 1.

      Listing of Common Variables with Conflicting Types
      Variable      Dataset           Type    Length
      height        WORK.COMPARE1     Char         8
                    WORK.COMPARE2     Num          8

     Listing of Common Variables with Differing Attributes
        Variable     Dataset           Type    Length  Format
        weight1      WORK.COMPARE1     Num          8   6.2
                     WORK.COMPARE2     Num          8

                     Observation Summary
               Observation      Base    Compare
               First Obs        1         1
               First Unequal    2         2
               Last Unequal     3         3
               Last Match       3         3
               Last Obs         4         .

Number of Observations in Common: 3.
Number of Observations in WORK.COMPARE1 but not in
WORK.COMPARE2: 1.
Total Number of Observations Read from WORK.COMPARE1: 4.
Total Number of Observations Read from WORK.COMPARE2: 3.


Number of Observations with Some Compared Variables Unequal: 2.
Number of Observations with All Compared Variables Equal: 1.
```

```
                        Values Comparison Summary
Number of Variables Compared with All Observations Equal: 1.
Number of Variables Compared with Some Observations Unequal: 2.
Total Number of Values which Compare Unequal: 2.
Maximum Difference: 0.1.
                        Variables with Unequal Values
Variable          Type    Len     Label                    Ndif    MaxDif
studentID         CHAR     8                                 1
weight1           NUM      8      first weight value         1      0.100

              Value Comparison Results for Variables
```

			Base Value	Compare Value		
	Obs	||	studentID	studentID		
	‾‾‾‾‾	||	‾‾‾‾‾‾‾	‾‾‾‾‾‾‾		
	3	||	A03	A04		

		first weight value			
		Base	Compare		
Obs	||	weight1	weight1	Diff.	% Diff
‾‾‾‾‾	||	‾‾‾‾‾‾‾	‾‾‾‾‾‾‾	‾‾‾‾‾‾	‾‾‾‾‾‾
2	||	150.20	150.3000	0.1000	0.0666

10.2.2 Comparing Observations with Common ID Values

In Program 10.3, values are compared based on the observations that occur in the same position. In this situation, you need to ensure the matching observations are from the same person or from the same sources. For example, the third observation in COMPARE1 has a value of "A03" for ID, whereas ID equals "A04" in the third observation of COMPARE2.

The results from the comparison in Program 10.3 might be just what you are looking for. In this situation, you can use the ID statement to confine your comparisons to those observations where the *ID* variable values match. To include the ID statement in PROC COMPARE, input data sets must be previously sorted by the named ID variable.

Program 10.4 uses STUDENTID in the ID statement to compare observations that have matching values for STUDENTID. Because the input data set has been created in STUDENTID order, there is no need to sort by STUDENTID again in the program.

Because the BY statement produces separate comparisons for male and female students, both data sets require a previous sort by the same BY variable. The VAR statement is used to restrict the comparison to the WEIGHT1 variable in both data sets.

By default, PROC COMPARE generates lengthy reports of comparison between the base and comparison data sets. The BRIEFSUMMARY (or BRIEF) option produces a short comparison between base and comparison data sets.

Program 10.4:

```
proc sort data = compare1;
    by gender;
run;

proc sort data = compare2;
    by gender;
run;

title 'PROC COMPARE: ID, BY and VAR statement';
proc compare base = compare1
             compare = compare2
             brief;
    by gender;
    id studentID;
    var weight1;
run;
```

Output from Program 10.4:

```
            PROC COMPARE: ID, BY and VAR statement
                   The COMPARE Procedure
         Comparison of WORK.COMPARE1 with WORK.COMPARE2
                      (Method = EXACT)

 – – – – – – – – – – – – - gender = F– – – – – – – – – – – –
NOTE: Data set WORK.COMPARE1 contains 1 observations not in
      WORK.COMPARE2.
NOTE: Values of the following 1 variables compare unequal:
      weight1

          Value Comparison Results for Variables
```

		first weight value			
	||	Base	Compare		
studentID	||	weight1	weight1	Diff.	% Diff
———	||	———	———	———	———
A04	||	149.20	149.1000	-0.1000	-0.0670

```
 – – – – – – – – – – – – - gender = M– – – – – – – – – – – –
NOTE: Values of the following 1 variables compare unequal:
      weight1
```

Value Comparison Results for Variables

studentID			first weight value Base weight1	Compare weight1	Diff.	% Diff
A02			150.20	150.3000	0.1000	0.0666

In the last two examples from this section, variables with the same name have been compared. However, it is possible to compare variables with different names.

Program 10.5 compares WEIGHT2 that is specified in the VAR statement from the *base* data set with WEIGHT1 specified by the WITH statement from the *comparison* data set. The ID statement also guarantees that the comparisons will be made between observations with matching values for STUDENTID.

Program 10.5:

```
title 'Compare variables with different variable names in
different data sets';
proc compare base = compare1
             compare = compare2
             brief;
    id studentID;
    var weight2;
    with weight1;
run;
```

Output from Program 10.5:

```
Compare variables with different variable names in different
data sets
                    The COMPARE Procedure
          Comparison of WORK.COMPARE1 with WORK.COMPARE2
                        (Method = EXACT)
NOTE: Data set WORK.COMPARE1 contains 1 observations not in
      WORK.COMPARE2.
NOTE: Values of the following 1 variables compare unequal:
      weight2^ = weight1
```

Value Comparison Results for Variables

studentID			Base weight2	Compare weight1	Diff.	% Diff
A04			149.2000	149.1000	-0.1000	-0.0670

| A01 | || 154.7000 | 154.3000 | -0.4000 | -0.2586 |
| A02 | || 150.2000 | 150.3000 | 0.1000 | 0.0666 |

You do not need the COMPARE= option to compare variables within the same data set. Program 10.6 compares two variables, WEIGHT1 and WEIGHT2, within the COMPARE1 data set.

Program 10.6:

```
title 'Compare variables with different variable names in the
same data sets';
proc compare base = compare1 brief;
    var weight1;
    with weight2;
run;
```

Output from Program 10.6:

```
Compare variables with different variable names in the
same data sets
                     The COMPARE Procedure
            Comparisons of variables in WORK.COMPARE1
                        (Method = EXACT)
NOTE: Values of the following 1 variables compare unequal:
      weight1^ = weight2
```

Valuc Comparison Results for Variables

			first weight value			
			Base	Compare		
Obs			weight1	weight2	Diff.	% Diff
——			———	———	———	———
1			149.10	149.2000	0.1000	0.0671
3			154.30	154.7000	0.4000	0.2592

10.3 Restructuring Data Sets Using the TRANSPOSE Procedure

Restructuring or transposing data sets were first introduced in Chapter 3 and Chapter 4 by using the DATA step programming. Chapter 6 introduced how to use array processing to efficiently transpose data. Another method for restructuring a data set is to use the TRANSPOSE procedure.

The material in this section is based upon a paper that I presented at SAS Global Forum (Li, 2012).

The syntax of PROC TRANSPOSE is listed below. The six statements in the TRANSPOSE procedure, which includes the PROC TRANPOSE, BY, COPY, ID, IDLABEL, and VAR statements, along with the eight options in the PROC TRANSPOSE statement, are used to apply different types of data transpositions and give the resulting data set a different appearance.

```
PROC TRANSPOSE  <DATA=input-data-set>
                <DELIMITER=delimiter>
                <LABEL=label>
                <LET>
                <NAME=name>
                <OUT=output-data-set>
                <PREFIX=prefix>
                <SUFFIX=suffix>;
    BY <DESCENDING> variable-1 <...<DESCENDING> variable-n>;
    COPY variable(s);
    ID variable;
    IDLABEL variable;
    VAR variable(s);
RUN;
```

10.3.1 Transposing an Entire Data Set

Program 10.7 starts by creating data set DAT1 with variables S_NAME, S_ID, E1, E2, and E3. The three numeric variables, E1 through E3, are labeled "English1", "English2", and "English3". By default, without specifying the names of the transposing variables, all the numeric variables from the input data set are automatically transposed.

In Program 10.7, the three numeric variables containing English test scores for the two students structured as a 2-by-3 matrix become three *observations* in a 3-by-2 matrix that defines the transposed data set for the two students newly represented as *variables*.

In the PROC TRANSPOSE statement, the OUT= option is used to specify the name of the transposed data set. Without an OUT= option, PROC TRANSPOSE will create a data set that uses the DATA*n* naming convention.

The NAME= option references the variable in the transposed data set that contains the names of the variables from the original data set which are being transposed. Without the NAME= option, the default _NAME_ is used.

Because E1 through E3 variables have permanent labels from the input data set, these labels are stored under the variable specified in the LABEL= option. Without specifying the LABEL = option, the default name _LABEL_ is used.

The PREFIX= option adds a prefix to the transposed variable names. Given PREFIX= score_, for example, the names of the transposed variables change to SCORE_1 and SCORE_2. Without the PREFIX= option, SAS uses the defaults, COL1 and COL2. You can also use the SUFFIX= option to attach a suffix to the transposed variable name.

Program 10.7:

```
data dat1;
    input s_name $ s_id $ e1 - e3;
    label e1 = English1
          e2 = English2
          e3 = English3;
    datalines;
John A01 89 90 92
Mary A02 92 .   81
;

proc transpose data = dat1
               out = dat1_out1
               name = varname
               label = labelname
               prefix = score_;
run;

title 'Dat1 in the original form';
proc print data = dat1 label;
run;

title 'Dat1 in transposed form';
proc print data = dat1_out1;
run;
```

Output from Program 10.7:

```
               Dat1 in the original form
    Obs    s_name    s_id    English1    English2    English3
     1     John      A01        89          90          92
     2     Mary      A02        92           .          81

               Dat1 in transposed form
    Obs    varname      labelname     score_1     score_2
     1       e1         English1        89          92
     2       e2         English2        90           .
     3       e3         English3        92          81
```

Program 10.8 adds more statements to PROC TRANSPOSE. The VAR statement is used to specify the variables that are to be transposed. In this instance, the VAR statement references the same E1 through E3 variables that are transposed by default. Thus, the output from Program 10.7 and Program 10.8 is the same.

The ID statement is used to specify the variable from the input data set that contains the values to rename the transposed variables. Because two variables S_NAME and S_ID are used in the ID statement along with the DELIM= option in the PROC TRANSPOSE statement, the values created by concatenating the S_NAME and the S_ID variables (separated by the value specified by the DELIM= option) are used as the names of the transposed variables.

The variable specified in the IDLABEL statement from the input data set can be either numeric or character and contains the values to label the transposed variables in the newly created transposed data set. From the partial output from the CONTENTS procedure, you can see that the names of the transposed variables are JOHN_A01 and MARY_A02, with A01 and A02 as their labels, respectively.

Program 10.8:

```
proc transpose data = dat1
               out = dat1_out2
               label = labelname
               name = varname
               delim = _;
    var e1-e3;
    id s_name s_id;
    idlabel s_id;
run;

title 'The use of VAR, ID, and IDLABEL statements';
proc print data = dat1_out2;
run;

proc contents data = dat1_out2;
run;
```

Output from Program 10.8:

```
      The use of VAR, ID, and IDLABEL statements
   Obs     varname     labelname     John_A01     Mary_A02
    1        e1        English1         89           92
    2        e2        English2         90            .
    3        e3        English3         92           81
```

Partial Output from PROC CONTENTS from Program 10.8:

```
       Alphabetic List of Variables and Attributes
    #     Variable      Type    Len     Label
    3     John_A01      Num      8      A01
    4     Mary_A02      Num      8      A02
    2     labelname     Char     40     LABEL OF FORMER VARIABLE
    1     varname       Char     8      NAME OF FORMER VARIABLE
```

10.3.2 Introduction to Transposing BY Groups

You can also transpose the data set BY group. More than one variable can be listed in the BY statement. To use the BY statement in PROC TRANSPOSE, the data set must be previously sorted using the same BY variable. When transposing a data set BY group, the BY variable is not transposed.

Program 10.9 transposes DAT1 by using S_NAME as the BY variable. The number of observations in the transposed data set (6) equals the number of BY groups (2) times the number of variables that are transposed (3). The number of transposed variables is equal to the number of observations within each BY group in the sorted input data set. For the input data processed in Program 10.9, the number of observations within each BY group is just one. Therefore, the number of transposed variables in the output data set is also just one (COL1).

The COPY statement in Program 10.9 copies the values from the S_ID variable from the input data set directly to the transposed data set. Because there are two observations from the input data set, the number of observations that will be copied will be two as well; SAS pads the missing values to the rest of the observations.

Program 10.9:

```
proc sort data = dat1 out = dat1_sort;
    by s_name;
run;

title 'DAT1_SORT where DAT1 is Sorted by S_NAME';
proc print data = dat1_Sort;
run;

proc transpose data = dat1_sort
               out = dat1_out3
               name = varname
               label = TEST;
    by s_name;
    copy s_id;
run;

title 'Transposing BY-group with the COPY statement';
proc print data = dat1_out3;
run;
```

Output from Program 10.9:

```
        DAT1_SORT where DAT1 is Sorted by S_NAME
    Obs    s_name    s_id    e1    e2    e3
     1     John      A01     89    90    92
     2     Mary      A02     92     .    81
```

```
          Transposing BY-group with the COPY statement
     Obs   s_name   s_id   varname     TEST      COL1
      1     John     A01      e1      English1    89
      2     John              e2      English2    90
      3     John              e3      English3    92
      4     Mary     A02      e1      English1    92
      5     Mary              e2      English2     .
      6     Mary              e3      English3    81
```

10.3.3 Where the ID Statement Does Not Work for Transposing BY Groups

You can use the ID statement to specify the variable from the input data set that contains the values to use for renaming transposed variables. In Program 10.9, the newly created transposed variable is given the uninformative name of COL1 by default. However, if you want to use the S_ID variable in the ID statement to change COL1, the results from Program 10.10 won't conform to your expectations. Now the transposed values in the output occupy *two* columns with A01 and A02 as variable names. What you expected is output where COL1 is simply renamed. To rename COL1, remove the ID statement and issue a RENAME suboption in the OUT= option attached to the PROC TRANSPOSE statement.

Program 10.10:

```
proc transpose data = dat1_sort
                out = dat1_out4
                name = varname
                label = TEST;
    by s_name;
    id s_id;
run;

title 'Incorrect way to use the ID statement';
proc print data = dat1_out4;
run;

proc transpose data = dat1_sort
                out = dat1_out5(rename = (col1 = Score))
                name = varname
                label = TEST;
    by s_name;
run;

title 'Rename COL1 to SCORE in the Output Data Set';
proc print data = dat1_out5;
run;
```

Output from Program 10.10:

```
Incorrect way to use the ID statement
Obs   s_name   varname     TEST      A01    A02
 1    John       e1      English1     89     .
 2    John       e2      English2     90     .
 3    John       e3      English3     92     .
 4    Mary       e1      English1      .     92
 5    Mary       e2      English2      .      .
 6    Mary       e3      English3      .     81

Rename COL1 to SCORE in the Output Data Set
Obs   s_name   varname     TEST      Score
 1    John       e1      English1     89
 2    John       e2      English2     90
 3    John       e3      English3     92
 4    Mary       e1      English1     92
 5    Mary       e2      English2      .
 6    Mary       e3      English3     81
```

10.3.4 Where the ID Statement Is Essential for Transposing BY Groups

Program 10.11 illustrates a situation where the ID statement is necessary in order to transpose data correctly. PROC TRANSPOSE in program 10.11 transposes one variable, SCORE, by using the variable S_NAME as the BY variable. The resulting transposed data set has two observations, which equals the number of BY groups (2) times the number of variables that are transposed (1). The problem with the transposed data set is that the third test score (81) for Mary is placed in the location reserved for the second test score.

Program 10.11:

```
data dat2;
    input s_name $ s_id $ exam score;
    datalines;
John A01 1 89
John A01 2 90
John A01 3 92
Mary A02 1 92
Mary A02 3 81
;

proc sort data = dat2 out = dat2_sort;
    by s_name;
run;

proc transpose data = dat2_sort out = dat2_out1;
    var score;
    by s_name;
run;
```

```
title 'Incorrect way to transpose - ID statement is not used';
proc print data = dat2_out1;
run;
```

Output from Program 10.11:

```
    Incorrect way to transpose - ID statement is not used
       Obs    s_name    _NAME_   COL1   COL2   COL3
        1      John      score     89     90     92
        2      Mary      score     92     81      .
```

Program 10.12 fixes the problem in Program 10.11 by using an ID statement along with the PREFIX = option. Both have to be used because ID references numeric variable EXAM. Without a user-defined PREFIX, the procedure would automatically insert an underscore before the numeric values in EXAM to create valid SAS variable names. Instead of the desired "test_1", you would get an inscrutable "_1".

Program 10.12:

```
proc transpose data = dat2_sort
               out = dat2_out2(drop = _name_)
               prefix = test_;
    var score;
    by s_name;
    id exam;
run;

title 'Correct way to transpose - ID statement is used';
proc print data = dat2_out2;
run;
```

Output from Program 10.12:

```
     Correct way to transpose - ID statement is used
        Obs     s_name    test_1    test_2    test_3
         1       John        89        90        92
         2       Mary        92         .        81
```

10.3.5 Handling Duplicated Observations Using the LET Option

John and Mary have two different scores for their third test in the DAT3 data set created in Program 10.13. If you use S_NAME as the BY variable and specify EXAM in the ID statement, PROC TRANSPOSE will not be able to transpose DAT3 and will generate the following error message written twice to the log:

```
ERROR: The ID value "test_3" occurs twice in the same BY group.
```

If the ID statement is removed from the code, PROC TRANSPOSE will be able to transpose DAT3, but the results will be sent to four columns, not three, because the maximum number of observations per BY group is four with the double entry for John.

For situations with duplicated records, you may want to keep only one record, such as keeping the largest or the smallest of the duplicated entries. The LET option from the PROC TRANSPOSE statement allows you to keep the last occurrence of a particular ID value within either the entire data set or a BY group.

Program 10.13 transposes DAT3 by keeping the largest value of each EXAM within each group of the S_NAME variable. Thus, it is necessary to sort the data by S_NAME first, followed by EXAM, and then finally by SCORE, in ascending order. Because of the sort, the highest score ends up being the last observation for each EXAM a student takes. Duplicate entries are not eliminated from the input data. Instead, SAS writes the same error message to the log; this time, however, the error is upgraded to a warning, so the data are successfully and accurately transposed.

Program 10.13:

```
data dat3;
    input s_name $ s_id $ exam score;
    datalines;
John A01 1 89
John A01 2 90
John A01 3 92
John A01 3 95
Mary A02 1 92
Mary A02 3 81
Mary A02 3 85
;

proc sort data = dat3 out = dat3_sort;
    by s_name exam score;
run;

proc transpose data = dat3_sort
                out = dat3_out(drop = _name_)
                prefix = test_
                let;
    var score;
    by s_name;
    id exam;
run;

title 'Keep the maximum score';
proc print data = dat3_out;
run;
```

Output from Program 10.13:

```
          Keep the maximum score
    Obs  s_name  test_1  test_2  test_3
     1    John     89      90      95
     2    Mary     92       .      85
```

Partial Log from Program 10.13:

```
69     proc transpose data = dat3_sort
70                    out = dat3_out(drop = _name_)
71                    prefix = test_
72                    let;
73        var score;
74        by s_name;
75        id exam;
76     run;
```

```
WARNING: The ID value "test_3" occurs twice in the same BY
         group.
NOTE: The above message was for the following BY group:
      s_name = John
WARNING: The ID value "test_3" occurs twice in the same BY
         group.
NOTE: The above message was for the following BY group:
      s_name = Mary
NOTE: There were 7 observations read from the data set
      WORK.DAT3_SORT.
NOTE: The data set WORK.DAT3_OUT has 2 observations and 4
      variables.
NOTE: PROCEDURE TRANSPOSE used (Total process time):
      real time      0.00 seconds
      cpu time       0.00 seconds
```

If you want to keep the smallest SCORE instead of the largest in the transposed data, all you need to do is sort S_NAME and EXAM in ascending order and then sort SCORE in descending order.

10.3.6 Situations Requiring Two or More Transpositions

In some applications, a single transposition will not produce the desired results. For example, to transpose DAT4 to DAT4_TRANSPOSE, you need to use PROC TRANSPOSE twice. Program 10.14A starts by creating the data set DAT4 and follows by transposing DAT4 by variable S_NAME.

DAT4:

	S_NAME	E1	E2	E3	M1	M2	M3
1	John	89	90	92	78	89	90
2	Mary	92	.	81	76	91	89

DAT4_TRANSPOSE:

	TEST_NUM	JOHN_E	JOHN_M	MARY_E	MARY_M
1	1	89	78	92	76
2	2	90	89	.	91
3	3	92	90	81	89

Program 10.14A:

```
data dat4;
   input s_name $ E1 - E3 M1 - M3;
datalines;
John 89 90 92 78 89 90
Mary 92 .  81 76 91 89
;

proc sort data = dat4 out = dat4_sort1;
   by s_name;
run;

proc transpose data = dat4_sort1 out = dat4_out1;
   by s_name;
run;

title 'First use of PROC TRANSPOSE for dat4';
proc print data = dat4_out1;
run;
```

Output from Program 10.14A:

```
              First use of PROC TRANSPOSE for dat4
                  Obs    s_name    _NAME_    COL1
                   1     John        E1       89
                   2     John        E2       90
                   3     John        E3       92
                   4     John        M1       78
                   5     John        M2       89
                   6     John        M3       90
                   7     Mary        E1       92
                   8     Mary        E2       .
                   9     Mary        E3       81
                  10     Mary        M1       76
                  11     Mary        M2       91
                  12     Mary        M3       89
```

Before performing a second transpose, you need to preprocess the output data set DAT4_OUT1 created in the first transposition. More specifically, two new variables have to be created: CLASS and TEST_NUM. CLASS is the first character in _NAME_, and TEST_NUM is the second and final character in _NAME_. Program 10.14B uses the SUBSTR function to create both TEST_NUM and CLASS.

Program 10.14B:

```
data dat4_out1a;
    set dat4_out1;
    test_num = substr(_name_,2);
    class = substr(_name_,1,1);
run;

title 'Creating TEST_NUM and CLASS variables';
proc print data = dat4_out1a;
run;
```

Output from Program 10.14B:

```
       Creating TEST_NUM and CLASS variables
   Obs   s_name   _NAME_   COL1   test_num   class
    1    John      E1       89       1         E
    2    John      E2       90       2         E
    3    John      E3       92       3         E
    4    John      M1       78       1         M
    5    John      M2       89       2         M
    6    John      M3       90       3         M
    7    Mary      E1       92       1         E
    8    Mary      E2        .       2         E
    9    Mary      E3       81       3         E
   10    Mary      M1       76       1         M
   11    Mary      M2       91       2         M
   12    Mary      M3       89       3         M
```

Program 10.14C sorts the data by TEST_NUM, within TEST_NUM by S_NAME, and finally within S_NAME by CLASS. Notice that the test scores in COL1 now have the desired order for the final transposition.

Program 10.14C:

```
proc sort data = dat4_out1a out = dat4_sort2;
    by test_num s_name class;
run;

title 'Sort data by TEST_NUM, S_NAME and CLASS';
proc print data = dat4_sort2;
run;
```

Output from Program 10.14C:

```
       Sort data by TEST_NUM, S_NAME and CLASS
   Obs    s_name   _NAME_   COL1   test_num   class
    1     John      E1       89       1         E
    2     John      M1       78       1         M
    3     Mary      E1       92       1         E
```

4	Mary	M1	76	1	M
5	John	E2	90	2	E
6	John	M2	89	2	M
7	Mary	E2	.	2	E
8	Mary	M2	91	2	M
9	John	E3	92	3	E
10	John	M3	90	3	M
11	Mary	E3	81	3	E
12	Mary	M3	89	3	M

PROC TRANSPOSE in Program 10.14D transposes COL1 by variable TEST_NUM and uses S_NAME and CLASS as the ID variables. The names of the transposed variables are separated by the underscore from the DELIMITER= option.

Program 10.14D:

```
proc transpose data = dat4_sort2
               out = dat4_out2(drop = _name_)
               delimiter = _;
   by test_num;
   var col1;
   id s_name class;
run;

title 'Second use of PROC TRANSPOSE for dat4';
proc print data = dat4_out2;
run;
```

Output from Program 10.14D:

```
       Second use of PROC TRANSPOSE for dat4
    Obs    test_num    John_E    John_M    Mary_E    Mary_M
     1        1          89        78        92        76
     2        2          90        89         .        91
     3        3          92        90        81        89
```

10.4 Creating the User-Defined Format Using the FORMAT Procedure

SAS provides a large selection of ready-made formats described earlier in Chapters 1 and 8. If you cannot find a format provided by SAS, you can create your own with the FORMAT procedure. Besides customized formats, PROC FORMAT can be used to create informats, list the contents of format catalogs, generate formats and informats from SAS data sets, and conversely create SAS data sets from formats or informats.

This section focuses on creating and printing customized formats and utilizing formats to create variables. Only the PROC FORMAT and the VALUE statements, listed in the syntax below, are presented in this section. Readers should refer to SAS documentation for other features of PROC FORMAT.

```
PROC FORMAT <LIBRARY=libref<.catalog>> < FMTLIB>;
   VALUE <$>name value-range-set(s);
RUN;
```

10.4.1 Creating User-Defined Formats

PROC FORMAT stores user-defined formats as entries in SAS catalogs. Without specifying the LIBRARY=*libref* option in the PROC FORMAT statement, newly created formats will be stored in the WORK.FORMATS catalog. WORK.FORMATS exists only for the duration of the current SAS session. If you specify *libref* without a *catalog* name, your formats will be stored in the *libref*.FORMATS catalog. For example, specifying the following statements, PROC FORMAT stores defined formats in the A.FORMATS catalog located in 'C:\' as FORMATS.SAS7BCAT.

```
libname a 'C:\';
proc format library = a;
```

Specifying LIBRARY=*libref.catalog-name*, the defined formats will be stored in the *libref.catalog-name* catalog. For example, specifying the following statements, PROC FORMAT stores defined formats in the A.MYFORMAT catalog located in 'C:\' as MYFORMAT.SAS7BCAT.

```
libname a 'C:\';
proc format library = a.myformat;
```

The VALUE statement is used to create a format that specifies character strings to use to print variable values. The *name* component in the VALUE statement is the name of the format that you are creating. Your format name cannot be the same as the name of a SAS-supplied format and must be a valid SAS name. Format names can have up to 32 characters. If a character format is being defined, the dollar sign ($) that appears in the first position counts as one of the 32 characters. Furthermore, a format name cannot end with a number because trailing numbers indicate lengths when formats are actually being applied. When creating a format in the VALUE statement, do not use a period after the format name.

The *value-range-set(s)* in the VALUE statement is used to specify one or more variable values and a character string. You can specify one or more *value-range-set(s)* in the following form:

```
value-or-range-1 <..., value-or-range-n>='formatted-value'
```

The *'formatted-value'* on the right side of the equal sign is used to specify a character string that becomes the printed value of the corresponding variable values listed as *value-or-range* sets that appears on the left side of the equal sign. The *'formatted-value'* must be character strings and can be up to 32,767 characters.

Based on the syntax above, you can specify multiple occurrences of *value-or-range*, separated by commas, in a value-range-set on the left side of the equal sign. Each occurrence of *value-or-range* can be specified as either *value* or *range*, which follows certain rules. Examples of specifying *value-or-range* are provided in Table 10.1.

The *value* in an occurrence of *value-or-range* is a single value, for example, 'A' or 1. For character formats, you need to enclose the character values, up to 32,767 characters, in single quotation marks. The *range* is used to specify a list of values, such as 5-10 (values from 5 to 10), or 'A'-'D' (values from 'A' to 'D').

The "less than" (<) symbol can be used to exclude values from ranges. If you are excluding the first value in a range, then you need to put the less than symbol (<) after the value. Similarly, if you are excluding the last value in a range, then put the < before the values. (See Example 3 in Table 10.1.)

If a value appears at the high end of one range as well as at the low end of another range, and the less than (<) symbol is not used, then PROC FORMAT assigns the value to the first range. (See Example 4 in Table 10.1.) If you overlap values in ranges, PROC FORMAT will issue an error message unless the MULTILABEL option is used (not covered in this book).

You can use the keywords LOW and HIGH as values that represent the lowest and highest numeric or character values in a range. The keyword LOW does not include numeric missing values, but it does include character missing values. You can also use the keyword OTHER to match all values that do not match any value or range or missing values. If LOW is used in a character range and OTHER is used as well, the missing value will not be matched in the OTHER group. (See Example 7 in Table 10.1.) Other examples of labeling missing character and numeric values and how the missing values are matched when the keywords OTHER and LOW are used are listed in Table 10.1.

Program 10.15 uses three FORMAT procedures and stores the $RACE, YESNO, $SMKFMT, SALFMT, and AGEFMT formats in three different catalogs. The first FORMAT procedure stores the $RACE YESNO and $SMKFMT formats in the FORMATS catalog in the WORK library. The $SMKFMT format is used for formatting values 'past' and 'never' with '0' and 'current' with '1'. Even though 0 and 1 are numeric values, they still need to be placed in quotation marks. The second FORMAT procedure stores the SALFMT format in the SALARY catalog in the DESKTOP library. The third FORMAT procedure stores the AGEFMT format in the FORMATS catalog in the DESKTOP library.

TABLE 10.1

Examples of Value-or-Range Sets

Example	All Possible Values	Value-or-Range	Values Being Labeled
1	2, 3, 4, 5, 6, 7, 8, and 9	`3, 5, 7, 9 = 'odd'` `2, 4, 6, 8 = 'even'`	3, 5, 7, and 9 are labeled with 'odd' 2, 4, 6, and 8 are labeled with 'even'

2	'A', 'B', 'C', 'D', and 'E'	`'A', 'C', 'E'`
		`= 'grp1'`
		`'B', 'D' = 'grp2'`
			'A', 'C', and 'E' are labeled with 'grp1'
			'B' and 'D' are labeled with 'grp2'

3	2, 3, 4, 5, 6, 7, 8, and 9	`2<-<6 = 'low'` `6-<9 = 'high'`	3, 4, and 5 are labeled with 'low' 6, 7, and 8 are labeled with 'high' 2 and 9 will be printed as 2 and 9 without any labels
4	2, 3, 4, 5, 6, 7, 8, and 9	`2-6 = 'low'` `6-9 = 'high'`	2, 3, 4, 5, and 6 are labeled with 'low' 7, 8, and 9 are labeled with 'high'

5	2, 3, 4, 5, 6, 7, 8, 9, and missing (.)	`LOW-6 = 'low'`
		`7-HIGH = 'high'`
		`OTHER = 'missing'`
			2, 3, 4, 5, and 6 are labeled with 'low'
			7, 8, and 9 are labeled with 'high'
			The missing value (.) is labeled with 'missing'

6	2, 3, 4, 5, 6, 7, 8, 9, and missing (.)	`LOW-6 = 'low'`
		`7-HIGH = 'high'`
		`. = 'missing'`
			2, 3, 4, 5, and 6 are labeled with 'low'
			7, 8, and 9 are labeled with 'high'
			The missing value (.) is labeled with 'missing'

7	'A', 'B', 'C', 'D', 'E', and missing (' ')	`LOW-'C' = 'low'`
		`'D'-HIGH = 'high'`
		`OTHER = 'missing'`
			'A', 'B', 'C', and missing value are labeled with 'low'
			'D' and 'E' are labeled with 'high'

8	'A', 'B', 'C', 'D', 'E', and missing (' ')	`'A'-'C' = 'low'`
		`'D'-HIGH = 'high'`
		`OTHER = 'missing'`
			'A', 'B', and 'C' are labeled with 'low'
			'D' and 'E' are labeled with 'high'
			The missing value is labeled with 'missing'

9	'A', 'B', 'C', 'D', 'E', and missing (' ')	`'A'-'C' = 'low'`
		`'D'-HIGH = 'high'`
		`' ' = 'missing'`
			'A', 'B', and 'C' are labeled with 'low'
			'D' and 'E' are labeled with 'high'
			The missing value is labeled with 'missing'

Program 10.15:

```
libname desktop 'C:\Documents and Settings\All Users\Desktop';
libname library 'C:\Documents and Settings\ARTHUR\Desktop\
   saslib';
proc format;
    value $race
        'A' = 'Asian'
        'B' = 'African American'
        'H' = 'Hispanic'
        'W' = 'White';
    value yesno
        1 = 'Yes'
        0 = 'No';
    value $smkfmt 'past', 'never' = '0'
        'current' = '1';
run;

proc format library = desktop.salary;
    value salfmt
        low - <50000 = 'low'
        50000 - <100000 = 'average'
        100000 - high = 'high';
run;

proc format library = desktop;
    value agefmt
        low - <30 = '< 30'
        30 - high = '> = 30'
        other = 'missing';
run;
```

Once the formats are created, you can use the FMTLIB option in the PROC FORMAT statement to display the contents of the format catalog. Program 10.16 prints the contents of the formats from the WORK.FORMAT, DESKTOP. FORMAT and DESKTOP.SALARY catalogs.

Program 10.16:

```
title 'Contents in WORK.FORMAT catalog';
proc format fmtlib;
run;

title 'Contents in DESKTOP.FORMAT catalog';
proc format library = desktop fmtlib;
run;

title 'Contents in DESKTOP.SALARY catalog';
proc format library = desktop.salary fmtlib;
run;
```

Output from Program 10.16:

```
                     Contents in WORK.FORMAT catalog
+- - - - - - - - - - - - - - - - - - - - - - - - - - - - - - +
| FORMAT NAME: YESNO LENGTH: 3 NUMBER OF VALUES: 2           |
| MIN LENGTH: 1 MAX LENGTH: 40 DEFAULT LENGTH 3 FUZZ: STD    |
+- - - - - - +- - - - - +- - - - - - - - - - - - - - - - - - +
|START       |END       |LABEL (VER. V7|V8 14AUG2012:11:06:48)|
+- - - - - - +- - - - - +- - - - - - - - - - - - - - - - - - +
|          0 |        0 |No                                  |
|          1 |        1 |Yes                                 |
+- - - - - - +- - - - - +- - - - - - - - - - - - - - - - - - +


+- - - - - - - - - - - - - - - - - - - - - - - - - - - - - - +
| FORMAT NAME: $RACE LENGTH: 16 NUMBER OF VALUES: 4          |
| MIN LENGTH: 1 MAX LENGTH: 40 DEFAULT LENGTH 16 FUZZ:  0    |
+- - - - - - +- - - - - +- - - - - - - - - - - - - - - - - - +
|START       |END       |LABEL (VER. V7|V8 14AUG2012:11:06:48)|
+- - - - - - +- - - - - +- - - - - - - - - - - - - - - - - - +
|A           |A         |Asian                               |
|B           |B         |African American                    |
|H           |H         |Hispanic                            |
|W           |W         |White                               |
+- - - - - - +- - - - - +- - - - - - - - - - - - - - - - - - +


+- - - - - - - - - - - - - - - - - - - - - - - - - - - - - - +
| FORMAT NAME: $SMKFMT LENGTH: 1 NUMBER OF VALUES: 3         |
| MIN LENGTH: 1 MAX LENGTH: 40 DEFAULT LENGTH 1 FUZZ:  0     |
+- - - - - - +- - - - - +- - - - - - - - - - - - - - - - - - +
|START       |END       |LABEL (VER. V7|V8 14AUG2012:11:06:48)|
+- - - - - - +- - - - - +- - - - - - - - - - - - - - - - - - +
|current     |current   |1                                   |
|never       |never     |0                                   |
|past        |past      |0                                   |
+- - - - - - +- - - - - +- - - - - - - - - - - - - - - - - - +


                   Contents in DESKTOP.FORMAT catalog
+- - - - - - - - - - - - - - - - - - - - - - - - - - - - - - +
| FORMAT NAME: AGEFMT LENGTH: 7 NUMBER OF VALUES: 3          |
| MIN LENGTH: 1 MAX LENGTH: 40 DEFAULT LENGTH 7 FUZZ: STD    |
+- - - - - - +- - - - - +- - - - - - - - - - - - - - - - - - +
|START       |END       |LABEL (VER. V7|V8 14AUG2012:09:26:30)|
+- - - - - - +- - - - - +- - - - - - - - - - - - - - - - - - +
|LOW         |         30<< 30                                |
|         30 |HIGH      |> = 30                              |
|**OTHER**   |**OTHER** |missing                             |
+- - - - - - +- - - - - +- - - - - - - - - - - - - - - - - - +
```

```
               Contents in DESKTOP.SALARY catalog
+- - - - - - - - - - - - - - - - - - - - - - - - - - - - +
| FORMAT NAME: SALFMT LENGTH: 7 NUMBER OF VALUES: 3      |
| MIN LENGTH: 1 MAX LENGTH: 40 DEFAULT LENGTH 7 FUZZ: STD |
+- - - - - +- - - - - +- - - - - - - - - - - - - - - - - +
|START     |END       |LABEL  (VER. V7|V8 14AUG2012:09:26:29)|
+- - - - - +- - - - - +- - - - - - - - - - - - - - - - - +
|LOW       |      50000<low                                |
|     50000|     100000<average                            |
|    100000|HIGH      |high                                |
+- - - - - +- - - - - +- - - - - - - - - - - - - - - - - +
```

10.4.2 Retrieving User-Defined Formats

Once formats are defined, you can associate them with the variables either permanently in the DATA step or temporarily in a PROC step. To retrieve a temporary format that is stored in the WORK library, you need to include the name of the format in the FORMAT statement because SAS automatically looks for the format in the WORK.FORMATS catalog. In addition, SAS automatically searches for formats in the LIBRARY.FORMATS. However, if LIBRARY is not used as the *libref* name or FORMATS is not specified as the catalog name, you must use the FMTSEARCH= system option and include the *libref* name or the catalog name or both in an OPTIONS statement; otherwise, SAS will not be able to locate your formats. Here is the syntax for specifying a list of format catalogs to search in the OPTIONS statement:

OPTIONS FMTSEARCH=(catalog-specification-1... catalog-
    specification-n);

Each *catalog-specification* can be *libref* or *libref.catalog*. If only *libref* is specified, then SAS assumes the catalog name is FORMATS. SAS searches the WORK. FORMATS catalog first, and then the LIBRARY.FORMATS catalog; SAS then searches the catalogs in the FMTSEARCH = list in the order in which they are listed until the format is found.

The HEARING data set used in Program 10.17 was created in Chapter 1. Additional information about the data set can be found in Table 1.1. Program 10.17 associates five user-defined formats, $RACE, YESNO, $SMKFMT, AGEFMT, and SALFMT, permanently to the variables RACE, PREG, SMOKE, AGE, and INCOME, respectively. Access to all formats is made possible by the FMTSEARCH option.

The two PRINT procedures in Program 10.17 show the impact that format assignments have upon printed output. PROC FREQ demonstrates how cross-tabulation of the two continuous variables AGE and INCOME is enhanced by using their formatted values.

Program 10.17:

```
options fmtsearch = (desktop desktop.salary);
data hearing2;
    set hearing;
    format race $race.
           preg yesno.
           smoke $smkfmt.
           age agefmt.
           income salfmt.;
run;

title 'Data set HEARING, formats are not used';
proc print data = hearing (obs = 5);
    var id race preg smoke age income;
run;

title 'Data set HEARING2, formats are used';
proc print data = hearing2 (obs = 5);
    var id race preg smoke age income;
run;

title 'Cross tabulation of AGE and INCOME by using formatted
value';
proc freq data = hearing2;
    tables age*income/missing;
run;
```

Output from Program 10.17:

```
            Data set HEARING, formats are not used
        Obs    id      race   Preg   smoke   Age    Income
         1    629F      H       0    past    26     35000
         2    656F      W       1    never   26     48000
         3    711F      W       1    never   32     30000
         4    511F      B       0    never   32     25000
         5    478F      W       0    past    34     35700

                Data set HEARING2, formats are used
    Obs    id      race               Preg  smoke   Age     Income
     1    629F   Hispanic              No      0   < 30      low
     2    656F   White                 Yes     0   < 30      low
     3    711F   White                 Yes     0   > = 30    low
     4    511F   African American      No      0   > = 30    low
     5    478F   White                 No      0   > = 30    low
```

```
Cross tabulation of AGE and INCOME by using formatted value
                   The FREQ Procedure
                 Table of Age by Income
```

```
Age          Income
Frequency|
Percent  |
Row Pct  |
Col Pct  |low      |average |high     | Total
- - - - -+- - - - +- - - - +- - - - +
missing  |       0 |       0 |       1 |      1
         |    0.00 |    0.00 |    2.94 |   2.94
         |    0.00 |    0.00 |  100.00 |
         |    0.00 |    0.00 |   33.33 |
- - - - -+- - - - +- - - - +- - - - +
< 30     |      14 |       6 |       1 |     21
         |   41.18 |   17.65 |    2.94 |  61.76
         |   66.67 |   28.57 |    4.76 |
         |   70.00 |   54.55 |   33.33 |
- - - - -+- - - - +- - - - +- - - - +
> = 30   |       6 |       5 |       1 |     12
         |   17.65 |   14.71 |    2.94 |  35.29
         |   50.00 |   41.67 |    8.33 |
         |   30.00 |   45.45 |   33.33 |
- - - - -+- - - - +- - - - +- - - - +
Total            20       11        3       34
              58.82    32.35     8.82   100.00
```

10.4.3 Creating Variables by Using User-Defined Formats

You can create categorical or indicator variables in a DATA step by exercising user-defined formats with PUT and INPUT functions. For example, Program 10.18 creates two variables, PREG_CHAR and SMOKE_IND, by using two of the formats created in Program 10.15.

Recall from Chapter 9 that the PUT function can convert either numeric values with numeric formats or character values with character formats to character values. Thus, the results generated from the PUT function are always characters. Because the PREG variable is numeric, the numeric format (YESNO) is used to convert PREG to the character variable PREG_CHAR.

To create the indicator variable SMOKE_IND, the PUT function converts the values from the SMOKE variable by using the $SMKFMT format to character values '1' and '0' first. Then the INPUT function converts the character results generated from the PUT function to numeric values by using the $w.$ informat.

Program 10.18:

```
data hearing3;
    set hearing;
    preg_char = put(preg, yesno.);
    smoke_ind = input(put(smoke, $smkfmt.), 1.);
run;
```

```
title 'Checking if PREG_CHAR and SMOKE_IND are created
correctly';
proc freq data = hearing3;
    tables preg*preg_char
           smoke*smoke_ind/nocol norow nopercent missing;
run;
```

Output from Program 10.18:

```
Checking if PREG_CHAR and SMOKE_IND are created correctly
                    The FREQ Procedure
                 Table of Preg by preg_char
      Preg         preg_char
      Frequency|      . |No       |Yes      | Total
      - - - -+- - - -+- - - -+- - - -+
            . |      4 |      0 |      0 |      4
      - - - -+- - - -+- - - -+- - - -+
            0 |      0 |     19 |      0 |     19
      - - - -+- - - -+- - - -+- - - -+
            1 |      0 |      0 |     11 |     11
      - - - -+- - - -+- - - -+- - - -+
      Total          4       19       11       34


                 Table of smoke by smoke_ind
      Smoke       smoke_ind
      Frequency|      . |      0 |      1 | Total
      - - - -+- - - -+- - - -+- - - -+
              |      1 |      0 |      0 |      1
      - - - -+- - - -+- - - -+- - - -+
      current |      0 |      0 |      8 |      8
      - - - -+- - - -+- - - -+- - - -+
      never   |      0 |     18 |      0 |     18
      - - - -+- - - -+- - - -+- - - -+
      past    |      0 |      7 |      0 |      7
      - - - -+- - - -+- - - -+- - - -+
      Total          1       25        8       34
```

10.5 Using the OPTIONS Procedure to Modify SAS System Options

In Section 10.4.2, the FMTSEARCH= system option is used in the OPTIONS statement to search user-defined formats. In addition, SAS provides numerous different types of system options to control how the output are formatted, how the files are being handled, or how the data sets are being processed, etc. To learn the current settings of a SAS system option, you can use the OPTIONS procedure, which has the following form:

```
PROC OPTIONS <LISTGROUPS><GROUP=><OPTION=>;
RUN;
```

Only a few options are listed in the syntax above. Other options can be found in SAS documentation. Without specifying any options, PROC OPTIONS will list all the system options in the SAS log. These system options can be categorized into different groups based on their functionality, and you can use LISTGROUPS to list the group names of all the system options. Once you know the name of each system option group, you can use the GROUP= option to display options belonging to one or more groups.

In Program 10.19, a few options are used in the OPTIONS statement to change the default system options. The NODATE option is used to suppress the date and time being printed in the listing output. NONUMBER is used to suppress the page number in the output. The LINESIZE= option sets the default line size to 65 characters per line in the SAS log and procedure output. The FORMDLIM= option specifies a blank line to delimit page breaks in the SAS output. All of these options are part of the LISTCONTROL system group and are listed by using the GROUP= option in the PROC OPTONS statement.

Program 10.19:

```
options nodate nonumber linesize = 65 formdlim = " ";
proc options group = listcontrol;
run;
```

Log from Program 10.19:

```
7    options nodate nonumber linesize = 65 formdlim = " ";
8    proc options group = listcontrol;
9    run;
   SAS (r) Proprietary Software Release 9.2 TS2M3

BYLINE                Print the byline at the beginning of each
                      by-group
CENTER                Center SAS procedure output
NODATE                Do not print date and time on top of each
                      page of SAS log and procedure output
NODETAILS             Do not display additional information in
                      directory lists
DMSOUTSIZE = 99999 Maximum number of rows in DMS output window
NODTRESET             Do not update date and time for log and print
                      output
FORMCHAR = ,ƒ"…†‡^‰Š‹Œ+ = |-/\<>*
                      Default output formatting characters for
                      print device
```

FORMDLIM =	**Character to delimit page breaks in SAS output**
FORMS = DEFAULT	Default form used to customize appearance of interactive windowing output
LABEL	Allow procedures to use labels with variables
LINESIZE = 65	**Line size for SAS log and SAS procedure output**
MISSING =.	Character to represent missing numeric value
NONUMBER	**Do not print page number on each page of SAS output**
NOPAGEBREAKINITIAL	
	Do not begin SAS log and listing files on a new page.
PAGENO = 1	Beginning page number for the next page of output produced by the SAS System
PAGESIZE = 56	Number of lines printed per page of output
NOPRINTINIT	Do not initialize SAS print file
SKIP = 0	Number of lines to skip before title
SYSPRINTFONT =	Set the default font for printing
HOSTPRINT	Print using the host Print Manager. To use Universal Printing or FORMS, this option must be turned off.
PRNGETLIST	Enable listing of the system printers. The SAS system will discover printers installed on the system.
SYSPRINT = ("Send To OneNote 2007")	
	Set the default printer and, optionally, an alternate destination (file) for output
NOTE: PROCEDURE OPTIONS used (Total process time):	
real time	0.00 seconds
cpu time	0.00 seconds

Another useful option of the PROC OPTIONS statement is the OPTION= option, which is used to display a short description and the value of specified options. For example, Program 10.20 lists the LINESIZE option in the log.

Program 10.20:

```
proc options option = linesize;
run;
```

Log from Program 10.20:

```
231  proc options option = linesize;
232  run;
   SAS (r) Proprietary Software Release 9.2 TS2M0
LINESIZE = 65      Line size for SAS log and SAS procedure
                   output
```

```
NOTE: PROCEDURE OPTIONS used (Total process time):
      real time             0.00 seconds
      cpu time              0.00 seconds
```

Exercises

Exercise 10.1. Program 10.14 transposes the DAT4 data set to the DAT4_ TRANSPOSE data set by using PROC TRANSPOSE twice. In this exercise, transpose the DAT4_TRANSPOSE data set back to the DAT4 data set.

Exercise 10.2. Based on the GRADE.SAS7BDAT data set, create three variables, MATH_POINT, ENGLISH_POINT, and PE_GRADE, using user-defined formats. The instructions for creating these three variables are described in *Exercise 2.1* in Chapter 2.

References

Cody, Ron. (2004). *SAS® Functions by Example*. Cary, NC: SAS Institute.

Cody, Ron. (2005). *Longitudinal Data and SAS® A Programmer's Guide*. Cary, NC: SAS Institute.

Cody, Ron, and Jeffrey Smith. (1991). *Applied Statistics and the SAS® Programming Language*, Fourth Edition. Upper Saddle River, NJ: Prentice Hall.

Li, A. (2008). "The Essence of DATA Step Programming," in *Proceedings of Western Users of SAS Software*, Universal Studio, CA.

Li, A. (2010). "Get the Scoop on the Loop: How Best to Write a Loop in the DATA Step," in *Proceedings of SAS Global Forum*, Seattle, WA.

Li, A. (2011). "The Many Ways to Effectively Utilize Array Processing," in *Proceedings of SAS Global Forum*, Las Vegas, NV.

Li, A. (2012). "Simplifying Effective Data Transformation via PROC TRANSPOSE," in *Proceedings of SAS Global Forum*, Orlando, FL.

SAS Institute. (2007). *SAS® Certification Prep Guide: Advanced Programming for SAS 9*. Cary, NC: SAS Institute.

SAS Institute. (2009). *Base SAS® 9.2 Procedures Guide*. Cary, NC: SAS Institute.

SAS Institute. (2009). *SAS® 9.2 Language Reference: Dictionary*, Second Edition. Cary, NC: SAS Institute.

SAS Institute. (2010). *SAS® 9.2 Language Reference: Concepts*, Second Edition. Cary, NC: SAS Institute.

Index

A

Alignment argument, 189
Ampersand format modifier, 168
Array processing, 121–137
   applications that use multi-
      dimensional arrays, 133–136
      calculating average SBP for
        pre- and post-treatment,
        133–134
      data set transformation, 136
      nested loop, 134
      number of observations, 135
      one-dimensional array, 134
      post-treatment results, 134
      pre-treatment measurements, 133
      restructuring data sets by using
        multi-dimensional array,
        135–136
      subscript, 133
      SUM statement, 135
      syntax, 133
      two-dimensional array, 134
   array applications, 130–133
      calculating products of multiple
        variables, 131–132
      creating a group of variables by
        using arrays, 130–131
      data set transformation, 132
      DO loop, 133
      list of variables, 130
      long format, 132
      pre-treatment measurements, 130
      product of multiple variables, 131
      restructuring data sets using
        one-dimensional arrays,
        132–133
      SUM function, 131
      temporary data elements, 131
   compilation and execution phases,
      125–126
      array name, 125
      iterations, 127, 128
      PDV creation, 125

      syntax errors, 126
   defining and referencing
      one-dimensional arrays,
      123–125
      array-elements, 123, 124
      array-name, 123, 124, 125
      asterisk, 123
      character elements, 123
      delimiters argument, 123
      initial-value-list, 124
      keywords, 124
      RETAIN statement, 124
      subscript, 125
   enriched syntax, 121
   exercises, 136–137
   functions and operators, 126–130
      array-elements, 126
      bound-n, 126
      DIM function, 127
      DIM, HBOUND, and LBOUND
        functions, 126–129
      functions, 126
      missing values, 129
      multi-dimensional array, 126
      OF operator, 130
      similar syntax, 126
      using IN and OF operator with
        array, 129–130
   situations for utilizing, 121–122
      DATA step variables, 122
      example, 121
      IF statements, 122
      multi-dimensional arrays, 122
      one-dimensional array, 122
      referencing of data, 122

B

BY-group processing in DATA step,
    79–93
   applications, 85–92
      calculating mean score within
        each BY group, 87–88
      counter variable, 87

creating data sets with duplicate or non-duplicate observations, 88–89
cumulating variable, 85, 87
data set observations, 92
example, 89
identical observations, 88
long format, 91
longitudinal data, 85
obtaining the most recent non-missing data within each BY group, 89–91
previously sorted variable, 87
restructuring data sets from long format to wide format, 91–92
RETAIN statement, 90
wide format, 91
BY-group processing, 79–86
execution phase of BY-group processing, 81–85
fifth iteration, 86
first iteration, 82
FIRST.VARIABLE and LAST.VARIABLE, 79–81
fourth iteration, 85
grouping variables, 79
longitudinal data, 79
multiple variables, 79
second iteration, 83
subsetting IF statement, 83
SUM statement, 81
third iteration, 84
total score for each subject, 81
exercises, 92–93

C

Cartesian product, 147
Catalog-specification, 233
Character variables, 2
Charlist argument, 195
Column pointer-control, 162
Combining data sets, 139–153
exercises, 152–153
horizontally combining data sets, 143–152
Cartesian product, 147
compilation phase, 144
descriptor information, 146, 149

example, 143
execution phase, 144, 146, 149
master data set, 151
match-merging, 147–151
merge-type, 145
merging syntax, 146
one-to-many matching, 147
one-to-one merging, 146
one-to-one reading, 143–145
output data set, 145
transaction data set, 151
unmatched observations, 150
updating data sets, 151–152
variables from different sources, 143
vertically combining data sets, 139–143
common variables, 141
compilation phase, 141, 143
concatenating data sets, 139–142
execution phase, 141, 143
input data sets, sorting of, 142
interleaving data sets, 142–143
new-name, 141
old-name, 141
program data vector, 141, 143
sum of observations, 139
syntax, 139
variables from different source, 139
COMPARE procedure, use of to compare contents of two data sets, 208–215
base data set, 214
common variables, 210
comparing observations with common ID values, 212–215
comparison data set, 214
first weight value, 209
information provided from PROC COMPARE, 209–212
matching variables, 209
observation differences, 210
options, 209
VAR statement, 212
Count argument, 194

D

Data entry error, 56
Data input and output, 155–179

creating text files, 175–177
    alignment parameters, 176
    column output, 175–176
    end-column, 175
    formatted output, 176
    last-column, 175
    list output, 177
    PUT statement, 177
    start-column, 175
    variable, 175
exercises, 177–179
reading text files, 160–174
    ampersand format modifier, 168
    character values, 161
    column input, 161–162
    column pointer-control, 162
    consecutive blanks, 168
    creating observations by using
        line-hold specifiers, 172–174
    creating observations by using
        line pointer-controls, 171–172
    data execution, 165
    data in free format, 169
    Delimiter-Sensitive-Data
        option, 167
    double trailing "at" signs, 172, 173
    embedded blanks, 168
    example, 163, 171
    formatted input, 162–164
    INFILE statement, 161, 172
    informat in modified list input, 169
    input buffer, 174
    input pointer, 171, 173
    INPUT statement, 164
    input values, 161
    line pointer-control, 162
    list input, 164–168
    missing values, 167
    mixed input, 170–171
    modified list input, 168–170
    nonstandard numeric value, 170
    pointer-control, 162
    program data vector, 160
    subsetting IF statement, 174
    syntax, modified list input, 168
reading and writing text files, 155–160
    character values, 157
    checklist, 156
    data format, 158

data informat, 157–158
data values, 159
decimal scaling factor, 158
embedded characters, 158
examples of date formats, 160
examples of date informats, 159
external-file, 155
FILENAME statement, 155
file-specification, 155, 156, 157
INFILE statement, 156
named input method, 156, 157
observations and variables, 155
SAS date and time values, 158–160
standard numeric values, 157
steps for reading text files,
    155–156
steps for writing text files, 157
text files, 156
Data sets, combining, *see* Combining
    data sets
Data step functions, 181–204
    character functions, 190–197
        assignment statement, 191
        CATT function, 192
        changing character cases, 190–191
        character-to-replace, 196
        character variable, 192
        charlist argument, 195
        concatenating character strings,
            191–194
        concatenation operator, 193
        count argument, 194
        default delimiter, 195
        delimiters argument, 190
        excerpt argument, 194
        functions for aligning character
            strings and trimming blanks,
            192, 193
        INDEX function, 195
        PROPCASE function, 190
        searching, exacting, and replacing
            character strings, 194–197
        source argument, 197
        string argument, 194
        syntax, 190
        target argument, 197
        TRANWARD function, 196
    date and time functions, 185–190
        alignment argument, 189

creating date and time values,
    185–186
date and time interval functions,
    188–190
date value, 185
DHMS function, 186
extracting components from date
    and time values, 187–188
functions for creating date and
    time values, 186
increment argument, 188
interval argument, 188
INTNX function, 189
numeric values, 185
start-from argument, 188, 189
variable description of TENANT
    data, 187
exercises, 203–204
field description for ID variable, 203
functions and CALL routines, 181–185
    algebraic expression, 182
    argument, 181, 183
    array-name, 181
    assignment statement, 181
    CALL routines, 182–184
    categories, 184–185
    character variables, 183
    descriptive statistics
        functions, 184
    functions, 181–182
    mathematical functions, 184
    MEAN function, 182
    numeric variables, 183
    RANUNI function, 185
    variable-list, 181
functions for converting variable
        types, 198–203
    automatic conversion, 199
    automatic numeric-to-character
        conversion, 201
    character informat, 200
    informat argument, 200
    INPUT function, 198–201
    missing value, 199
    operator, 201
    PUT function, 201–203
    returned values, 202
    reversed direction of
        conversion, 198

source argument, 200, 201
DATA step operation, 55–77
    conditional processing in DATA step,
        66–70
    detecting the end of data set by
        using the END= option, 68
    expression, 67
    final data set, 70
    initialized variable, 68
    long format, 68
    observations, 69
    restructuring data sets from wide
        format to long format, 68–70
    subsetting IF statement, 66–67
    temporary variable, 68
    wide format, 68
DATA step processing overview,
        55–62
    compilation phase, 55
    data entry error, 56
    DATA step compilation phase,
        57–58
    DATA step execution phase, 58–61
    declarative statements, 55
    difference between reading a raw
        data set and SAS data set,
        61–62
    end-of-file marker, 59
    error, 55
    executable statements, 55
    execution phase, 55
    first iteration, 59
    importance of OUTPUT
        statement, 61
    iteration, 55
    program data vector, 57
    raw data file, 61
    second iteration, 60
    syntax errors, 58
    third iteration, 60
debugging techniques, 70–76
    BREAK command, 75
    format option, 74
    highlighted statement, 74
    logic errors, 70, 74
    missing variable, 76
    strategy, 70
    syntax errors, 70
    using DATA step debugger, 74–76

using PUT statement to observe
contents of PDV, 70–73
variable-list, 71
exercises, 76–77
retaining the value of newly created
variables, 62–66
declarative statement, 63
execution phase, 63
expression, 65
first iteration, 64
initialized variable, 62
RETAIN statement, 62–64
second iteration, 65
SUM statement, 64–66
third iteration, 66
Date values, 159
Decimal scaling factor, 158
Declarative statement, 55, 63
Delimiters argument, 123, 190
Descriptive statistics functions, 184

E

End-of-file marker, 59
Excerpt argument, 194
Executable statements, 55
Exercises
array processing, 136–137
BY-group processing in DATA step,
92–93
combining data sets, 152–153
conditional creation of variables, 54
data input and output, 177–179
data step functions, 203–204
DATA step operation, 76–77
introduction to SAS®, 33–34
SAS® procedures, 239
writing loops in DATA step, 117–119
Explicit loops, 96–102
CENTER variable, 103
combining loops, 103
DO loop, first two iterations, 99
expression, evaluated, 101
identical statements, 97
ID variable, 101
increment value, 98
index-variable, 98, 102
inner loop, 102
last two iterations, 100

nested loops, 102
number of iterations, 99
outer loop, 102
programming error, 101
start value, 98
stop value, 98

F

File-specification, 111, 155, 156, 157
Formatted input method, 164

G

Global statements, 1
Grouping variables, 79

H

Horizontally combining data sets,
143–152
Cartesian product, 147
compilation phase, 144
descriptor information, 146, 149
example, 143
execution phase, 144, 146, 149
master data set, 151
match-merging, 147–151
merge-type, 145
merging syntax, 146
one-to-many matching, 147
one-to-one merging, 146
one-to-one reading, 143–145
output data set, 145
transaction data set, 151
unmatched observations, 150
updating data sets, 151–152
variables from different sources, 143

I

IF-THEN/ELSE statement, 35–45
bytes of storage space, 42
character variable, comparison of, 38
DO group, 43–45
evaluation of variables, 35
FREQ procedure, 43
handling missing values when
creating variables, 37–39
indicator variable, 36

LENGTH attribute, 41–43
  multiple, 45–48, 53
  numeric variable, 40
  PROC MEANS, 39
  steps for creating a variable, 35–37
  TRUE and FALSE (logical
      expressions), 39–41
  variable attributes, 35
Implicit loops, 95–96
  DATA step execution, 96
  example, 95
  implicit OUTPUT statement, 96
  RANUNI function, 96
Increment argument, 188
Increment value, 98
Index-variable, 98, 102
Informat argument, 200
Input and output, *see* Data input
      and output
Interval argument, 188

L

Line pointer-control, 162
List input method, 166
List output method, 177
Longitudinal data, 79, 85
Loops, *see* Writing loops in DATA step

M

Master data set, 151
Match-merging, 147
Mathematical functions, 184

N

Named input method, 156, 157
Nested loops, 102, 134
Numeric variables, 2

O

OPTIONS procedure, use of to modify
      SAS system options, 236–239
  LINESIZE option, 238
  listing output, 237
  options, 237, 238
  page breaks, 237
  settings, 237

P

PDV, *see* Program data vector
Pointer-control, 162
Procedures, *see* SAS® procedures
Program data vector (PDV), 57, 141, 160

R

RACE variable, 30
Reading text files, 160–174
  ampersand format modifier, 168
  character values, 161
  column input, 161–162
  column pointer-control, 162
  consecutive blanks, 168
  creating observations by using
      line-hold specifiers, 172–174
  creating observations by using line
      pointer-controls, 171–172
  data execution, 165
  data in free format, 169
  Delimiter-Sensitive-Data option, 167
  double trailing "at" signs, 172, 173
  embedded blanks, 168
  example, 163, 171
  formatted input, 162–164
  INFILE statement, 161, 172
  informat in modified list input, 169
  input buffer, 174
  input pointer, 171, 173
  INPUT statement, 164
  input values, 161
  line pointer-control, 162
  list input, 164–168
  missing values, 167
  mixed input, 170–171
  modified list input, 168–170
  nonstandard numeric value, 170
  pointer-control, 162
  program data vector, 160
  subsetting IF statement, 174
  syntax, modified list input, 168

S

SAS®, introduction to, 1–34
  base SAS procedures, 13–24
    BY variable, 14

CLASS statement, 18
common statements in SAS
    procedures (TITLE, BY, and
    WHERE statements), 13–14
CONTENTS procedure, 14–16
default title, 13
descriptor, 14
FREQ procedure, 20–24
HEARING data set, 15
MEANS procedure, 18–20
PRINT procedure, 16–18
RACE variable, 17
SMOKE variable, 19
SORT procedure, 16
TABLES statement, 20
VARNUM option, 15
VAR statement, 17
WORK library, 14
changing the appearance of data,
    28–33
category of ethnicity, 30
character data, 31
dissociation of format from
    variable, 32
formatting variable values using
    SAS FORMATS, 31–33
labeling variables, 29–31
nonstandard numeric format, 32
preferred format, 28
RACE variable, 30
removal of labels from
    variables, 29
creating and modifying variables,
    9–13
assignment statement and SAS
    expression, 9–12
creating variables conditionally,
    12–13
expression categories, 11
function, 11
IF-THEN/ELSE statements, 9, 12
INCOME variable, 13
mnemonic-equivalent forms, 9
PRINT procedure, 12
exercises, 33–34
reading data into SAS, 2–9
character variables, 2
floating-point numbers, 2
INFILE statement, 7

input data set, 4
input methods, 7
library, 2
message component, 4
naming rules, 3
numeric variables, 2
output data set, 4
reading data entered directly into
    program, 8–9
reading a raw data file with fixed
    fields, 6–8
reading a SAS data set, 3–6
rows, 2
SAS data set and SAS library, 2–3
SAS log, 5
terminologies, 2
text file, 6
SAS program and language, 1
global statements, 1
language elements, 1
PROC steps, 1
subsetting data by selecting
    variables, 24–28
DROP= data set option or DROP
    statement, 27
KEEP= data set option or KEEP
    statement, 25–26
name prefix lists, 24
numbered range lists, 24
variable-list notation, 25
variable-lists, types of, 24
where to specify DROP= and
    KEEP= data set options and
    DROP/KEEP statements, 27–28
SAS® procedures, 205–239
creating user-defined format using
    FORMAT procedure, 227–236
catalog-specification, 233
continuous variables, 233
creating user-defined formats,
    228–233
creating variables by using
    user-defined formats, 235–236
examples of value-or-range
    sets, 230
format catalog, 231
formatted-value, 229
HEARING data set, 233
indicator variable, 235

informats, 227
keywords, 229
retrieving user-defined formats,
    233–235
value-or-range, 229
value-range-set, 228
VALUE statement, 228
WORK library, 233
exercises, 239
restructuring data sets using
    TRANSPOSE procedure,
    215–227
CONTENTS procedure, 218
COPY statement, 219
duplicated records, 223
handling duplicated observations
    using the LET option, 222–224
introduction to transposing BY
    groups, 219–220
observations, 216
RENAME suboption, 220
situations requiring two or more
    transpositions, 224–227
suffix, 217
syntax, 216
transposing entire data set,
    216–218
uninformative name, 220
user-defined PREFIX, 222
variables, 216
where ID statement does not work
    for transposing BY groups,
    220–221
where ID statement is essential
    for transposing BY groups,
    221–222
using COMPARE procedure to
    compare contents of two data
    sets, 208–215
base data set, 214
common variables, 210
comparing observations with
    common ID values, 212–215
comparison data set, 214
first weight value, 209
information provided from PROC
    COMPARE, 209–212
matching variables, 209
observation differences, 210

options, 209
VAR statement, 212
using OPTIONS procedure to
    modify SAS system options,
    236–239
LINESIZE option, 238
listing output, 237
options, 237, 238
page breaks, 237
settings, 237
using SORT procedure to eliminate
    duplicate observations,
    205–208
adjacent observations, 207
data set, 205
eliminating duplicate
    observations, 207–208
eliminating observations with
    duplicate BY values, 205–207
input data set, 206
NODUPKEY, 205, 206
observations, 205
PROC SORT, 206
SORT procedure, 205
Select-expression, 48, 49
SORT procedure, use of to eliminate
    duplicate observations,
    205–208
adjacent observations, 207
data set, 205
eliminating duplicate observations,
    207–208
eliminating observations with
    duplicate BY values, 205–207
input data set, 206
NODUPKEY, 205, 206
observations, 205
PROC SORT, 206
SORT procedure, 205
Source argument, 197, 200, 201
Start-from argument, 188, 189
String argument, 194
Syntax
array processing, 121, 126
character functions, 190
combining data sets, 139, 146
errors, DATA step operation, 58, 70
reading text files, 168
TRANSPOSE procedure, 216

T

Target argument, 197
Text files, *see* Data input and output
Transaction data set, 151
TRANSPOSE procedure, restructuring
 data sets using, 215–227
 CONTENTS procedure, 218
 COPY statement, 219
 duplicated records, 223
 handling duplicated observations
  using the LET option, 222–224
 introduction to transposing BY
  groups, 219–220
 observations, 216
 RENAME suboption, 220
 situations requiring two or more
  transpositions, 224–227
 suffix, 217
 syntax, 216
 transposing entire data set, 216–218
 uninformative name, 220
 user-defined PREFIX, 222
 variables, 216
 where ID statement does not work
  for transposing BY groups,
  220–221
 where ID statement is essential
  for transposing BY groups,
  221–222

U

User-defined format, creation of
 using FORMAT procedure,
 227–236
 catalog-specification, 233
 continuous variables, 233
 creating user-defined formats,
  228–233
 creating variables by using user-
  defined formats, 235–236
 examples of value-or-range sets, 230
 format catalog, 231
 formatted-value, 229
 HEARING data set, 233
 indicator variable, 235
 informats, 227
 keywords, 229

 retrieving user-defined formats,
  233–235
 value-or-range, 229
 value-range-set, 228
 VALUE statement, 228
 WORK library, 233

V

Variables
 character, 2
 continuous, 233
 counter, 87
 creating and modifying, 9–13
  assignment statement and SAS
   expression, 9–12
  creating variables conditionally,
   12–13
  expression categories, 11
  function, 11
  IF-THEN/ELSE statements, 9, 12
  INCOME variable, 13
  mnemonic-equivalent forms, 9
  PRINT procedure, 12
 cumulating, 85, 87
 dissociation of format from, 32
 grouping, 79
 indicator, 235
 -lists, types of, 24
 matching, 209
 numeric, 2
 product of, 131
 removal of labels from, 29
 temporary, 104
Variables, conditional creation of,
 35–54
 executing one of several statements,
  45–52
  AGEGROUP variable, 45
  executing statements using
   SELECT group, 48–52
  multiple IF-THEN/ELSE
   statements, 45–48
  OTHERWISE statement, 49
  select-expression, 48, 49
  threshold value, 46
  when-expression, 49
 exercises, 54
 IF-THEN/ELSE statement, 35–45

bytes of storage space, 42
character variable, comparison
    of, 38
DO group, 43–45
evaluation of variables, 35
FREQ procedure, 43
handling missing values when
    creating variables, 37–39
indicator variable, 36
LENGTH attribute, 41–43
numeric variable, 40
PROC MEANS, 39
steps for creating a variable, 35–37
TRUE and FALSE (logical
    expressions), 39–41
variable attributes, 35
modifying of IF-THEN/ELSE
    statement with assignment
    statement, 52–54
    AGEGROUP variable, 53, 54
    comparison evaluation, 53
    indicator variable, 52
    multiple IF-THEN/ELSE
        statement, 53
Vertically combining data sets, 139–143
    common variables, 141
    compilation phase, 141, 143
    concatenating data sets, 139–142
    execution phase, 141, 143
    input data sets, sorting of, 142
    interleaving data sets, 142–143
    new-name, 141
    old-name, 141
    program data vector, 141, 143
    sum of observations, 139
    syntax, 139
    variables from different source, 139

W

When-expression, 49
WORK library, 14
Writing loops in DATA step, 95–119
    exercises, 117–119
    implicit and explicit loops, 95–103
        CENTER variable, 103
        combining loops, 103
        DATA step execution, 96

DO loop, first two iterations, 99
example, 95
explicit loops, 96–102
expression, evaluated, 101
identical statements, 97
ID variable, 101
implicit loops, 95–96
implicit OUTPUT statement, 96
increment value, 98
index-variable, 98, 102
inner loop, 102
last two iterations, 100
nested loops, 102
number of iterations, 99
outer loop, 102
programming error, 101
RANUNI function, 96
start value, 98
stop value, 98
using looping to read a list of
    external files, 110–117
    DO loop, first iteration, 114
    DO loop, second iteration, 115
    DO loop, third iteration, 116
    DO UNTIL loop, 111
    dummy, 112
    end-of-file marker, 110, 111
    explicit OUTPUT statement, 113
    external file, 110–111
    file-specification, 111
    INFILE statement, 111
    multiple external files, 111–117
    placeholder, 111
    reading multiple files, 112
    specifying a condition, 110
    variable, 110
utilizing loops to create samples,
    103–109
    algorithm, 108, 109
    components, 104
    CONTENTS procedure, 105
    creating random sample with
        replacement, 106–108
    creating random sample without
        replacement, 108–109
    creating systematic sample,
        105–106
    data set, 103

direct-access mode, 104–105
DO loop, first iteration, 106
increment values, 105
input-data-set, 105
last two iterations, 107

observation to be selected, 108
Rannum, 108
sampling schemes, 103
STOP statement, 104
temporary variable, 104